Techniques and Topics in
FLOW MEASUREMENT

Frank E. Jones

CRC Press
Boca Raton   New York   London   Tokyo

Library of Congress Cataloging-in-Publication Data

Jones, Frank E.
    Techniques and topics in flow measurement / by Frank E. Jones.
        p.   cm.
    Includes bibliographical references and index.
    ISBN 0-8493-2475-0 (alk. paper)
    1. Fluid dynamic measurements. I. Title.
TA357.5.M43J66 1995
    620.1′.064′0287—dc20   95-16808
                                                                                CIP

No claim to original U.S. Government works
International Standard Book Number 0-8493-2475-0
Library of Congress Card Number 95-16808
Printed in the United States of America 1 2 3 4 5 6 7 8 9 0
Printed on acid-free paper

THE AUTHOR

Frank E. Jones is currently an independent consultant. He received his bachelor's degree in physics from Waynesburg College, Pennsylvania, and his master's degree in physics from the University of Maryland, where he has also pursued doctoral studies in meteorology. He served as a physicist at the National Bureau of Standards (now the National Institute of Standards and Technology, NIST) in many areas including flow measurements, standardization for chemical warfare agents, chemical engineering, processing of nuclear materials, nuclear safeguards, evaporation of water, humidity sensing, evapotranspiration, cloud physics, earthquake research, mass, length, time, volume, and sound.

Mr. Jones began work as an independent consultant upon his retirement from NIST in 1987. He is the author of more than 75 technical publications, three published books, with another book in preparation, and he also holds two patents. He is an Associate Editor of the National Council of Standards Laboratories Newsletter. He is a member of the American Society for Testing and Materials, the Instrument Society of America, and the Institute for Nuclear Materials Management, and is associated with other technical societies from time to time as they are relevant to his interests.

*It is a privilege to dedicate this book
to the late great statistician,
personality and friend, Churchill Eisenhart.*

He that believeth on me, as the scripture hath said, out of his belly shall flow rivers of living water.

—*John 7:38*

PREFACE

There is available a great deal of valuable and useful literature on flow measurement, including at least one excellent handbook. It is not the intent of the present book to cover, in many cases redundantly, the entire field of flow measurement; therefore, many areas will not be found here. It is, rather, the intent here to present many items (which I trust will be of value and useful) from my own work on flow measurement and related topics, some of which have not been published in the open literature and some of which have been published in journals. Therefore, this book would be supplementary to other books and to other published works. I have attempted to present material that is current and can be useful to specialists in flow measurement and to those whose involvement in the field is incidental to other work or interests.

The book begins with a discussion of three systems of units: SI, the English Absolute dimensional system, and the English Engineering system. The relationships between the base units of mass, length, and time are given, as are the relations between the three systems of units as they apply to various quantities such as force, weight, and pressure. An application is made to manometry. Eleven exercises are included.

Chapters on the density of gases and the density of water follow. In the first of these two chapters, the air density equation is developed in detail. Also, equations that can be used to calculate the density of seven other gases are presented. Tables of the composition of dry air, the compressibility factor for air, and the saturation vapor pressure of water and nine exercises are included. In the chapter on the density of water, new equations for the calculation of the density of water (air-free and air-saturated) are developed using values from the work of Kell. Tables of the density of air-free water, the density of air-saturated water, and (for comparison) the density of air-free water calculated using the Wagenbreth and Blanke formulation are provided. One exercise is included.

In Chapter 4, the water vapor mixing ratio (the ratio of the mass of water in a moist air mixture to the mass of dry air in the mixture) as a flow parameter is discussed. Examples of the relationships between the mixing ratio and other flow parameters are given. One exercise is included.

Simple interpolation formulas relating the viscosity of a gas to its temperature and pressure are developed in Chapter 5. The formulas are fitted to the excellent accurate, precise, and internally consistent experimental data of Kestin and co-workers. Tables of calculated values of viscosity for dry air, nitrogen, carbon dioxide, helium, argon, and oxygen are provided. Six examples are included.

In Chapter 6, equations are developed for the calculation of the ratio of the specific heat at constant pressure to the specific heat at constant volume, gamma, for dry air, water vapor, and moist air. Two tables and three examples are included.

In 1992, the National Institute of Standards and Technology (NIST) and the international standards community instituted a new policy on expressing measurement uncertainty. The policy groups the components of uncertainty in the result of a measurement in two categories: category A (type A)—those components that are evaluated by statistical methods—and category B (type B)—those components that are evaluated by other than statistical methods. Chapter 7 discusses the components of uncertainty, standard uncertainty, type A evaluation of standard uncertainty, type B evaluation of standard uncertainty, combined standard uncertainty, expanded uncertainty, and relative uncertainties. The application of NIST guidelines is illustrated with three examples.

For accurate weighing, corrections accounting for buoyancy forces must be applied. In Chapter 8, buoyancy corrections are discussed and applied to flow calibration. Three exercises are included.

An interpolation formula to be used in the calculation of the real-gas critical flow factor is developed in Chapter 9. Two exercises are included. Subsonic flow of gas through venturis, nozzles, and orifices is briefly discussed in Chapter 10. Expressions for the nozzle discharge coefficient and the discharge coefficient for orifice plates are derived.

The precision with which automatic pipets can be used to dispense known volumes of water in the volume calibration of tanks is illustrated in Chapter 11.

A new reference method for the testing of hydrometers is presented in Chapter 12. The method is simple and easy to use and has a precision of approximately 0.01%.

A new treatment of calibration data for laminar flowmeters is presented in Chapter 13. A corrected form of the Poiseuille equation is developed, and a set of calibration data for dried air, nitrogen, helium, and argon is used to determine the values of the coefficients in the equation.

A new treatment of the effect of kinematic viscosity on the performance of turbine flowmeters for liquids is presented in Chapter 14. The "linearization" process previously described by Jones delineates the region of linear performance of the turbine flowmeter and determines the parameters in a linear equation relating volume flow rate to frequency or pulse rate. The effect of kinematic viscosity is assigned to the parameters. Calibration data for three turbine flowmeters for a number of values of kinematic viscosity are used to illustrate the treatment, which is superior to the usual "universal viscosity curve" approach in which the "K-factor" is involved.

A preliminary investigation of calibration data for vortex-shedding flowmeters is presented in Chapter 15. Calibration data for two universal venturi meters are fitted to linear equations relating discharge coefficient to Reynolds number in Chapter 16.

A new treatment of calibration data to ascertain the linear range of anemometers is presented in Chapter 17. The treatment is similar to the "linearization" process previously described by Jones for turbine flowmeters.

A new method for estimating the correction (diverter correction) to be made to the measured time interval for diverters is presented in Chapter 18. Data for two diverters are used to illustrate the method.

The calibration of a platform scale used to determine the mass of water in a tank used for flowmeter calibration is illustrated in Chapter 19. Calibration of volumetric test measures is discussed in Chapter 20.

The determination of liquid density by hydrostatic weighing and by a mechanical oscillator technique is discussed in Chapter 21.

The volume calibration of tanks which can be used to dispense liquid through a flowmeter for calibration is discussed in Chapter 22. The method, which was refined by Jones for volume calibration of chemical process tanks, relates height (and therefore volume) of the liquid to differential pressure between the inside bottom of the tank and surface of the liquid.

Frank E. Jones
Potomac, Maryland
January 1995

TABLE OF CONTENTS

Chapter 1

FORCE, WEIGHT, AND PRESSURE UNITS

INTRODUCTION

The only internationally recognized system of units is the SI (Le Systeme International d'Unites), with the base units mass = kilogram (kg), length = meter (m), and time = second (s). However, the English Absolute dimensional system and the English Engineering system are still in use. Therefore, the specialist in flow measurement, and engineers in general, should be familiar with the relations[1] between these three systems of units as they apply to various quantities such as force, weight, and pressure.

FORCE

Newton's second law of motion (F = ma) *defines* force (F) in terms of the units of mass (m) and acceleration (a). Note that acceleration is a derived unit (m/s^2).

$$F = ma \tag{1.1}$$

In SI units

$$1 \text{ newton} = 1 \text{ kg} \times 1 \text{ m/s}^2 \tag{1.2}$$

In English Absolute units

$$1 \text{ poundal} = 1 \text{ lb}_m \times 1 \text{ ft/s}^2 \tag{1.3}$$

In English Engineering units

$$1 \text{ lb}_f = 1 \text{ slug} \times 1 \text{ ft/s}^2 \tag{1.4}$$

or

$$1 \text{ lb}_f = 1 \text{ lb}_m \times 1 \text{ g}_0 \tag{1.5}$$

where g_0 = standard acceleration due to gravity = 9.80665 m/s^2 = 32.17405 ft/s^2.
  We see from the above that

1. A *force* of 1 newton (N) accelerates a *mass* of 1 kg by 1 m/s^2.
2. A *force* of 1 poundal accelerates a *mass* of 1 pound (lb_m) by 1 ft/s^2.
3. A *force* of 1 pound (lb_f) accelerates a *mass* of 1 pound (lb_m) by 32.17405 ft/s^2, or accelerates a *mass* of 1 slug by 1 ft/s^2.

1

These are definitions, not derivations.

All of these equations are the same, differing only in the units used for force and mass:

Force — newton (N), poundal, or lb-force (lb_f)
Mass — kg, lb-mass (lb_m), or slug

These simple equations *define* the units and their relationships.

We can relate the three systems of units by using a constant, k, with Newton's second law of motion:

$$F = kma \qquad\qquad (1.6)$$

where k is a dimensionless proportionality constant that can be used to rectify the units for different systems. Of course, in SI, when mass is in kilograms, length is in meters, time is in seconds, F is in newtons (N), where 1 newton is equal to 1 kg m/s², and k is unity. A force of 1 N imparts to a body of mass 1 kg an acceleration of 1 m/s² or an acceleration of 1 m/s² of a body of mass 1 kg results in (or from) a force of 1 N. The magnitude of k is thus unity.

In the English Absolute dimensional system (1), the unit of mass is the pound (lb_m), also referred to as pound-mass, the unit of length is the foot (ft), and the unit of time is the second (s). The corresponding unit of force is the *poundal* (lb_m ft/s²), that is, a force of 1 poundal imparts to a body of mass 1 lb_m an acceleration of 1 ft/s², or an acceleration of 1 ft/s² of a body of mass 1 lb_m results in (or from) a force of 1 poundal. Again, the magnitude of k is unity.

In the English Engineering system, however, the unit of force is the *pound-force* (lb_f), the unit of mass is the pound (lb_m), the unit of length is the foot (ft), and the unit of time is the second (s). The subscripts on "lb" are a consequence of the (unfortunate) choice of the term "pound" for both the force unit and the mass unit. Note that the "pound" (avoirdupois pound) is a unit of *mass*, defined in terms of the kilogram by

$$1 \text{ avoirdupois pound} = 0.45359237 \text{ kilogram} \qquad\qquad (1.7)$$

The pound-force is *defined* as that force that will impart to a body of mass 1 lb_m an acceleration of 1 g_0 or 32.17405 ft/s².

The "standard" acceleration due to gravity (g_0) is defined as 9.80665 m/s², which is equal to 32.17405 ft/s². If the acceleration in the definition of pound-force, g_0, is expressed in English units, then

$$k = F/ma = 1 \text{ } lb_f/(1 \text{ } lb_m \times 32.17405 \text{ ft/s}^2) \qquad\qquad (1.8)$$

We can now define g_c (as commonly used in engineering) by the following:

$$g_0 = g_c \times 1 \text{ ft/s}^2 \qquad\qquad (1.9)$$

$$g_c = g_0/(1 \text{ ft/s}^2) \tag{1.10}$$

$$g_c = (32.17405 \text{ ft/s}^2)/(1 \text{ ft/s}^2) \tag{1.11}$$

$$g_c = 32.17405$$

Thus, we can say that, in English Engineering units,

$$k = 1 \text{ lb}_f/(1 \text{ lb}_m \times g_c \times 1 \text{ ft/s}^2) \tag{1.12}$$

Numerically, $k = 1/g_c$; g_c is said to have "units" of $(\text{lb}_m \text{ ft})/(\text{lb}_f \text{ s}^2)$.

It is unfortunate that the *number* 32.17405 is given the symbol g_c, since gravitational acceleration is conventionally represented by the symbol "g" with dimensions LT^{-2} and units of m/s^2 or ft/s^2.

Complication arises in the British Engineering case because the proportionality constant k is not unity or, stated another way, the force unit is defined as that associated with the acceleration of unit mass by $g_c \times 1 \text{ ft/s}^2$.

Note the relationship between the pound-force and the poundal:

$$1 \text{ lb}_f = 1 \text{ lb}_m \times 32.17405 \text{ ft/s}^2 \tag{1.13}$$

$$1 \text{ poundal} = 1 \text{ lb}_m \times 1 \text{ ft/s}^2 \tag{1.14}$$

$$(1 \text{ lb}_f/1 \text{ poundal}) = 32.14705 = g_c \tag{1.15}$$

Therefore, $1 \text{ lb}_f = 32.17405$ poundals. We can thus consider the pound-force to be a force quantity expressed in terms of the poundal. Similarly, the slug is a mass quantity — $1 \text{ slug} = 32.17405 \text{ lb}_m$.

WEIGHING

Weighing is the process by which the mass of an object is determined by the gravitational force on it, corrected for the buoyant force exerted on the object by the medium (usually air) in which the object is weighed. For simplicity, we shall assume that weighings are done in vacuum and then apply buoyancy corrections later.

In a gravitational field, a body is acted on by gravitational force, F_g, due to gravitational attraction. On an equal-arm balance, two objects of equal mass balance each other if the gravitational forces are the same. The two objects will balance each other independent of the value of g_L, the local acceleration due to gravity, because the effect of g_L is the same on each.

On a spring balance, however, the gravitational force on an object is balanced by the restoring force of a spring. As the value of g_L changes from location to location, the gravitational force on the object changes and the indication of the spring balance changes. It is necessary, therefore, to adjust the spring balance to indicate mass at the location of the balance (at the value of g_L at the location). If, for example, the balance were adjusted to indicate mass at a location at which the acceleration due to gravity is equal to g_0 and if the balance were used at a location at which the acceleration of gravity is equal to a different value, g_L, the ratio of the indicated mass to the mass of the body would be g_L/g_0.

The gravitational force in the English Absolute system is

$$F_g(\text{poundals}) = m(\text{lb}_m) \times g_L(\text{ft/s}^2) \qquad (1.16)$$

The gravitational force in the English Engineering system is

$$F_g(\text{lb}_f) = (g_L/g_c)(\text{ft/s}^2) \times m(\text{lb}_m) \qquad (1.17)$$

Recalling that g_c is the *magnitude* (32.17405) of g_0, at a location for which $g_L = g_0$, the gravitational *force* on an object in lb_f is *numerically equal* to the *mass* of the object in lb_m. A spring balance would, when calibrated and adjusted with objects of known mass, thus indicate the *mass* of the object. At a location for which g_L is not equal to g_0, the scale would require adjustment to compensate for the ratio g_L/g_0.

To account for the buoyant force in weighing, the gravitational force is multiplied by the factor $(1 - \rho_a/\rho_m)$, where ρ_a is the density of the air in which the weighing is made and ρ_m is the density of the object being weighed. A complete discussion of buoyancy will be given in Chapter 8.

PRESSURE

In SI, the unit of pressure is the pascal (Pa):

$$1 \text{ Pa} = 1 \text{ N/m}^2 = 1 \text{ kg/(m s}^2) \qquad (1.18)$$

In the English Absolute system, the unit of pressure is the poundal/ft²:

$$1 \text{ poundal/ft}^2 = 1 \text{ lb}_m/(\text{ft s}^2) \qquad (1.19)$$

In the English Engineering system, the unit of pressure is the pound-force/ft² or the pound-force/in²:

$$1 \text{ lb}_f/\text{ft}^2 = 32.17405 \text{ lb}_m/(\text{ft s}^2) \qquad (1.20)$$

$$1 \text{ lb}_f/\text{in}^2 = 1 \text{ lb}_f/(\text{ft}^2/144) = 144 \text{ lb}_f/\text{ft}^2 \qquad (1.21)$$

$$1 \text{ lb}_f/\text{in}^2 = 144 \times 32.17405 \text{ lb}_m/(\text{ft s}^2) \qquad (1.22)$$

$$= 4633.063 \text{ lb}_m/(\text{ft s}^2) \qquad (1.23)$$

The lb_f/in^2 is abbreviated "PSI".

$$1 \text{ PSI} = 4633.063 \text{ } lb_m/(ft \text{ } s^2) \quad (1.24)$$

$$1 \text{ PSI} = 4633.063 \times (0.45359237 \text{ kg})/(0.3048 \text{ m } s^2) \quad (1.25)$$

$$1 \text{ PSI} = 6894.757 \text{ kg}/(m \text{ } s^2) = 6894.757 \text{ N}/m^2 \quad (1.26)$$

$$1 \text{ PSI} = 6894.757 \text{ Pa} \quad (1.27)$$

The pressure due to 1 atmosphere is 101325 Pa. The same pressure in PSI is 101325 Pa/[6894.757(Pa/PSI)], which is equal to 14.69595 PSI. Thus,

$$1 \text{ atmosphere} = 101325 \text{ Pa} = 14.69595 \text{ PSI} \quad (1.28)$$

APPLICATION TO MANOMETRY

The hydrostatic equation is of the form

$$dP = \rho \text{ } g_L \text{ } dh \quad (1.29)$$

where dP is a change in pressure, ρ is the density of the fluid, g_L is the local acceleration due to gravity, and dh is a change in height.

In SI, dP is in Pa, ρ is in kg/m^3, g_L is in m/s^2, and dh is in m.

In the English Absolute system, dP is in $poundal/ft^2$, ρ is in lb_m/ft^3, g_L is in ft/s^2, and dh is in ft.

In the English Engineering system, the hydrostatic equation is

$$dP = \rho \text{ } (g_L/g_c) \text{ } dh \quad (1.30)$$

where dP is in lb_f/ft^2, ρ is in lb_m/ft^3, g_L is in ft/s^2, g_c is the *number* 32.17405, and dh is in ft.

We now consider a water manometer with a fluid of density ρ_s in contact with water in both legs (Figure 1.1). The pressure on the high-pressure side of the manometer is P_2; the pressure on the low-pressure side is P_1; the height of water in the low-pressure leg is H_2; the height of water in the high-pressure leg is H_1; the density of water in the manometer is ρ_w; the density of the manometer fluid is ρ_s; and the local acceleration due to gravity is g_L.

The pressure at the height H_2 in the high-pressure leg is equal to $P_2 + dP_s$; the pressure at the height H_2 in the low-pressure leg is equal to $P_1 + dP_s$. dP_s is the differential pressure due to the column of fluid above the height H_2. The pressure at the surface of the water at the height H_1 in the high-pressure leg is equal to $P_2 + dP_s + dP_{s2}$, where dP_{s2} is the differential pressure due to the column of fluid of height $H_2 - H_1$.

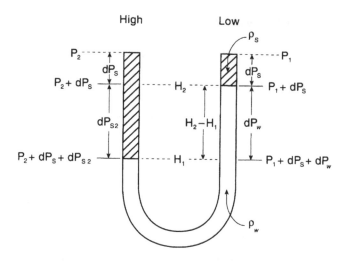

FIGURE 1.1. Water manometer with fluid of density ρ_s in contact with water in both legs. *(From Measurements and Control, June 1989, with permission.)*

The pressure at the height H_1 in the low-pressure leg is equal to $P_1 + dP_s + dP_w$, where dP_w is the differential pressure due to the column of water of height $H_2 - H_1$. The pressure at the height H_1 in each leg is the same, thus

$$P_2 + dP_s + dP_{s2} = P_1 + dP_s + dP_w \qquad (1.31)$$

$$P_2 - P_1 = dP_w - dP_{s2} \qquad (1.32)$$

By substitution, we arrive at

$$dP_w = \rho_w(g_L/g_c)(H_2 - H_1) \qquad (1.33)$$

$$dP_{s2} = \rho_s(g_L/g_c)(H_2 - H_1) \qquad (1.34)$$

Thus,

$$P_2 - P_1 = (g_L/g_c)(r_w - r_s)(H_2 - H_1) \qquad (1.35)$$

This simple example illustrates the occurrence of g_c in a case in which pressure is expressed in lb_f/ft^2. Again, it is emphasized that g_c is the *number* 32.17405.

EXERCISES

1.   Express the acceleration due to gravity of 32.14703 ft/s² in m/s². What percentage is this of the standard acceleration due to gravity?

2. Express a mass of 175 lb_m in slugs and in kg.
3. When one steps on a scale, is the indication of the scale the mass of the individual or the force exerted by that mass?
4. Express a force of 118 poundals in lb_f and in N.
5. Express the pressure 14.63 PSI in atmospheres and in Pa.
6. Express the standard acceleration due to gravity in ft/s^2.
7. Express the standard acceleration due to gravity in m/s^2.

REFERENCE

1. **Jones, F. E.,** Force, weight, and pressure units, *Measurements Control*, June, 178, 1989.

Chapter 2

DENSITY OF GASES

INTRODUCTION

The density of gases is often required in the various calculations involved in flow measurements. In this chapter, the air density equation will be developed in detail. Also, equations for the calculation of the density of other gases of interest will be presented.

DEVELOPMENT OF THE AIR DENSITY EQUATION[1]

The ideal gas equation,

$$PV = nRT \tag{2.1}$$

relates the total pressure, P, the total volume, V, and the absolute temperature, T, of an ideal gas or a mixture of ideal gases. The number of moles of the gas or the mixture of gases is n and R is the universal gas constant or molar gas constant.

In terms of density, ρ, rather than volume, Equation 2.1 becomes

$$P = \rho RT/M \tag{2.2}$$

where M is the molecular weight of the gas or the apparent molecular weight of the mixture.

For a mixture of dry air (indicated by the subscript a) and water vapor (indicated by the subscript w), ρ and M are, respectively, the density and apparent molecular weight of the air-water vapor mixture. Because

$$M = m/n = (m_a + m_w)/(n_a + n_w) \tag{2.3}$$

where m is the mass of the mixture and n is the number of moles of the mixture,

$$M = (n_a M_a + n_w M_w)/(n_a + n_w) = M_a \left[(1 + n_w M_w/n_a M_a)/(1 + n_w/n_a) \right] \tag{2.4}$$

We now introduce the water vapor mixing ratio, r:

$$r = (\text{mass of water vapor/mass of dry air}) = n_w M_w/n_a M_a \tag{2.5}$$

and designate the ratio M_w/M_a by ϵ, whereby Equation 2.4 becomes

$$M = M_a (1 + r)/(1 + r/\epsilon) \tag{2.6}$$

9

We now substitute Equation 2.6 in Equation 2.2 and note that the effective water vapor pressure, e', in moist air is defined[2] by

$$e' = rP/(\epsilon + r) \qquad (2.7)$$

then

$$P = (\rho RT/M_a)\{1/[1 + (\epsilon - 1)e'/P]\} \qquad (2.8)$$

Equation 2.8 is the *ideal* gas equation for a mixture of dry air and water vapor with water vapor pressure of e'. If the air-water vapor mixture behaved as a mixture of ideal gases,

$$P/(\rho RT/M_a)\{1/[1 + (\epsilon - 1)e'/P]\} = Z = 1 \qquad (2.9)$$

where Z is the compressibility factor.

Since a mixture of air and water vapor is not ideal, the magnitude of the nonideality is reflected in the departure of Z from 1. Equation 2.9 then becomes

$$P = (\rho RTZ/M_a)\{1/[1 + (\epsilon - 1)e'/P]\} \qquad (2.10)$$

Equation 2.10 is the *real* gas equation for a mixture of dry air and water vapor. By rearrangement of Equation 2.10, the expression for the air density is

$$\rho = (PM_a/RTZ)[1 + (\epsilon - 1)e'/P] \qquad (2.11)$$

PARAMETERS IN THE AIR DENSITY EQUATION

UNIVERSAL GAS CONSTANT, R

The value of the molar gas constant, R, listed in a compilation by Cohen and Taylor,[3] is 8.31441 ± 0.00026 J K^{-1} mol^{-1}. Quinn et al.[4] made a new determination of R by measuring the speed of sound in argon by means of an acoustic interferometer. Their value was 8315.73 ± 0.17 J K^{-1} kmol^{-1}. Gammon[5] deduced a value of R from measurements of the speed of sound in helium; his later reported value[6] is 8315.31 ± 0.35 J K^{-1} kmol^{-1}, which, considering the uncertainties, was in close agreement with the value of Quinn et al. Rowlinson and Tildesley[7] interpreted the experimental measurements of Quinn et al. and arrived at a value of 8314.8 ± 0.3 J K^{-1} kmol^{-1}, which is in close agreement with the value of Cohen and Taylor within the uncertainties attached to the values.

Jones[1] chose to use the value of Cohen and Taylor "with the realization that in the future it might be replaced by a new value." Such a new value has appeared. Moldover et al.[8,9] reported a new experimental value of the gas constant of 8.314471 ± 0.000014 J mol^{-1} K^{-1} where the uncertainty quoted is a standard deviation. This uncertainty is smaller by a factor of 5 than that of earlier values. The value of Moldover et al. is the preferred value and will be used throughout this chapter.

APPARENT MOLECULAR WEIGHT OF AIR, M_a

The apparent molecular weight of dry air, M_a, is calculated as a summation, Σ, using the relationship

$$M_a = \Sigma_i^k \, M_i x_i \qquad (2.12)$$

where each M_i is the molecular weight of an individual constituent and x_i is the corresponding mole fraction, the ratio of the number of moles of a constituent to the total number of moles in the mixture.

The molecular weights and typical mole fractions of the constituents of dry air are tabulated in Table 2.1. Other constituents are present in abundances which are negligible for the present application.

The values of the atomic weights of the elements[10] are based on the carbon-12 scale. The molecular weights are taken to be the sums of the atomic weights of the appropriate elements.

The value for the abundance of oxygen is taken from Reference 11. The value for the abundance of carbon dioxide is taken from an unpublilshed compilation of data on *atmospheric* concentration of carbon dioxide at seven locations throughout the world. It must be emphasized that 0.00033 was the mole fraction of CO_2 in the atmosphere and should be considered a *background* value. The mole fraction of CO_2 in laboratories, which is of course the value of interest here, is in general greater than 0.00033 and is *variable*. For example, three samples of air taken from a glove box in the Mass Laboratory at the National Institute of Standards and Technology (NIST) had a mean value of 0.00043, and four samples of laboratory air taken at the National Center for

TABLE 2.1
Composition of Dry Air

Constituent	Abundance (mole fraction)	Molecular Weight
Nitrogen	0.78102	28.0134
Oxygen	0.20946	31.9988
Carbon dioxide	0.00033	44.0098
Argon	0.00916	39.948
Neon	0.00001818	20.179
Helium	0.00000524	4.00260
Krypton	0.00000114	83.80
Xenon	0.000000087	131.30
Hydrogen	0.0000005	2.0158
Methane	0.0000015	16.0426
Nitrous oxide	0.0000003	44.0128

Atmospheric Research in Boulder, CO, had a mean value of 0.00080. Clearly, then, the optimum utilization of the air density calculation would necessitate a measurement of CO_2 abundance on an air sample taken at the time of the laboratory measurements of interest.

One of the options one has in dealing with the variability of CO_2 abundance is to select a reference level, e.g., 0.00033 or 0.00043, and to provide an adjustment to M_a to account for known departures from the reference level.

Gluekauf,[12] in discussing the variation of the abundance of oxygen in the atmosphere, stated that "all major variations of the O_2 content must result from the combustion of fuel, from the respiratory exchange of organisms, or from the assimilation of CO_2 in plants. The first process does not result in more than local changes of O_2 content, while the latter two processes, though locally altering the CO_2/O_2 ratio, leave their sum unchanged."

The constancy of the sum is expressed by the equation (for convenience, the subscript i has been replaced by the chemical symbol):

$$x_{CO_2} + x_{O_2} = \text{constant} = 0.20979 \tag{2.13}$$

The contribution of O_2 and CO_2 to the apparent molecular weight of dry air is

$$M_{O_2}x_{O_2} + M_{CO_2}x_{CO_2} = 31.9988 \, x_{O_2} + 44.0098 \, x_{CO_2} \tag{2.14}$$

From Equation (2.13),

$$x_{O_2} = 0.20979 - x_{CO_2} \tag{2.15}$$

and

$$M_{O_2}x_{O_2} + M_{CO_2}x_{CO_2} = 12.011 \, x_{CO_2} + 6.7130 \tag{2.16}$$

Therefore,

$$\delta \, (M_a) = \delta \, [M_{O_2}x_{O_2} + M_{CO_2}x_{CO_2}]$$

$$= 12.011 \, \delta \, (x_{CO_2}) \tag{2.17}$$

that is, the variation in M_a due to a variation in CO_2 abundance is equal to 12.011 (the atomic weight of carbon) multiplied by the variation in CO_2 abundance.

The variation in M_a due to the difference between the reference levels 0.00033 and 0.00043 is thus 0.0012 g mol^{-1}, which corresponds to a relative variation of 41 ppm (parts per million) in M_a and a corresponding variation of 41 ppm in air density.

The adjusted M_a accounting for the departure of the CO_2 abundance from the reference level of 0.00033 becomes

$$M_a = M_{a033} + 12.011 \, [x_{CO_2} - 0.00033] \qquad (2.18)$$

where M_{a033} is the apparent molecular weight of dry air with a CO_2 mole fraction of 0.00033.

The value of the abundance of argon in dry air, 0.00916, is that calculated from the mass spectrometric determination of the ratio of argon to argon and nitrogen by Hughes.[13]

The value for the abundance of nitrogen was arrived at by the usual practice of inferring nitrogen abundance to be the difference between unity and the sum of the mole fractions of the other constituents.

The abundances of the constituents neon through nitrous oxide in Table 2.1 were taken to be equal to the parts per volume concentration in U.S. Standard Atmosphere, 1976.[14]

From the data of Table 2.1, the apparent molecular weight of dry air with a CO_2 mole fraction of 0.00033 is calculated by Equation 2.12 to be 28.963. For dry air with a CO_2 mole fraction of 0.00043, the apparent molecular weight is calculated to be 28.964.

COMPRESSIBILITY FACTOR, Z

The compressibility factor was computed using the virial equation of state of an air-water vapor mixture expressed as a power series in reciprocal molar volume,

$$Z = Pv/RT = 1 + B_{mix}/v + C_{mix}/v^2 + \ldots \qquad (2.19)$$

and expressed as a power series in pressure,

$$Z = Pv/RT = 1 + B'_{mix}P + C'_{mix}P^2 + \ldots \qquad (2.20)$$

where v is the molar volume, B_{mix} and B'_{mix} are second virial coefficients, and C_{mix} and C'_{mix} are third virial coefficients for the mixture.

The virial coefficients of the pressure series are related to the virial coefficients of the volume power series by

$$B'_{mix} = B_{mix}/RT \qquad (2.21)$$

and

$$C'_{mix} = (C_{mix} - B^2_{mix})/(RT)^2 \qquad (2.22)$$

Each mixture virial coefficient is a function of the mole fractions of the individual constituents and the virial coefficients for the constituents. The virial coefficients are functions of temperature only.

Using the virial coefficients[15] provided by Hyland, a table of compressibility factor, Z, for CO_2-free air, Table 2.2, has been generated. Table 2.2 is applicable to moist air containing reasonable amounts of CO_2.

Alternatively, Z can be calculated using the following equations:
For P in pascals and t in °C,

$$Z = 1.00001 - 5.8057 \times 10^{-9}\,P + 2.6402 \times 10^{-16}\,P^2$$

$$- 3.3297 \times 10^{-7}\,t + 1.2420 \times 10^{-10}\,Pt$$

$$- 2.0158 \times 10^{-18}\,P^2 t + 2.4925 \times 10^{-9}\,t^2$$

$$- 6.2873 \times 10^{-13}\,Pt^2 + 5.4174 \times 10^{-21}\,P^2 t^2$$

$$- 3.5 \times 10^{-7}\,(RH) - 5.0 \times 10^{-9}\,(RH)^2 \qquad (2.23)$$

For P in PSI (pounds per square inch) and t in °C,

$$Z = 1.00001 - 4.0029 \times 10^{-5}\,P + 1.2551 \times 10^{-8}\,P^2$$

$$- 3.3297 \times 10^{-7}\,t + 8.5633 \times 10^{-7}\,Pt$$

$$- 9.5826 \times 10^{-11}\,P^2 t + 2.4925 \times 10^{-9}\,t^2$$

$$- 4.3349 \times 10^{-9}\,Pt^2 + 2.5753 \times 10^{-13}\,P^2 t^2$$

$$- 3.5 \times 10^{-7}\,(RH) - 5.0 \times 10^{-9}\,(RH)^2 \qquad (2.24)$$

For temperatures and/or pressures outside the range of Table 2.2, the table of compressibility factor of moist air (also CO_2-free) in the Smithsonian Meteorological Tables[16] can be used, with some loss of precision since the listing there is to the fourth decimal place.

RATIO OF THE MOLECULAR WEIGHT OF WATER TO THE MOLECULAR WEIGHT OF DRY AIR, ϵ

The molecular weight of water is 18.0152.[10] The ratio, ϵ, of the molecular weight of water to that of dry air is, therefore, 0.62201 for dry air with a CO_2 mole fraction of 0.00033. For dry air with a CO_2 mole fraction of 0.00043, the ratio, ϵ, is 0.62199.

EFFECTIVE WATER VAPOR PRESSURE, e'

Because e' is the effective vapor pressure of water in *moist air*, a word of caution with regard to inferring e' from measurements of relative humidity is in order. Relative humidity, U, can be defined[17] by

$$U = (e'/e_s') \times 100 \text{ percent} \qquad (2.25)$$

where e_s' is the *effective* saturation vapor pressure of water in moist air. The introduction of a second gas (air in this case) over the surface of water increases

TABLE 2.2
Compressibility Factor, Z, Calculated Using Equation 2.23

t (°C)	P (Pa)	RH (%)				
		0	25	50	75	100
19.0	95000	0.99964	0.99963	0.99961	0.99960	0.99957
	100000	0.99962	0.99961	0.99959	0.99958	0.99956
	101325	0.99962	0.99960	0.99959	0.99957	0.99955
	105000	0.99960	0.99959	0.99958	0.99956	0.99954
	110000	0.99989	0.99957	0.99956	0.99954	0.99952
20.0	95000	0.99965	0.99964	0.99962	0.99960	0.99958
	100000	0.99963	0.99962	0.99960	0.99958	0.99956
	101325	0.99963	0.99962	0.99960	0.99958	0.99956
	105000	0.99961	0.99960	0.99958	0.99957	0.99954
	110000	0.99959	0.99958	0.99957	0.99955	0.99953
21.0	95000	0.99966	0.99965	0.99963	0.99961	0.99958
	100000	0.99964	0.99963	0.99961	0.99959	0.99956
	101325	0.99964	0.99962	0.99961	0.99959	0.99956
	105000	0.99962	0.99961	0.99559	0.99957	0.99955
	110000	0.99960	0.99959	0.99958	0.99956	0.99953
22.0	95000	0.99967	0.99965	0.99963	0.99961	0.99958
	100000	0.99965	0.99964	0.99962	0.99960	0.99957
	101325	0.99965	0.99963	0.99961	0.99959	0.99956
	105000	0.99963	0.99962	0.99960	0.99958	0.99955
	110000	0.99962	0.99960	0.99958	0.99956	0.99954
23.0	95000	0.99968	0.99966	0.99964	0.99962	0.99959
	100000	0.99966	0.99964	0.99962	0.99960	0.99957
	101325	0.99965	0.99964	0.99962	0.99960	0.99957
	105000	0.99964	0.99963	0.99961	0.99958	0.99956
	110000	0.99963	0.99961	0.99959	0.99957	0.99954
24.0	95000	0.99968	0.99967	0.99965	0.99962	0.99959
	100000	0.99967	0.99965	0.99963	0.99961	0.99958
	101325	0.99966	0.99965	0.99963	0.99960	0.99957
	105000	0.99965	0.99964	0.99962	0.99959	0.99956
	110000	0.99964	0.99962	0.99960	0.99957	0.99954
25.0	95000	0.99969	0.99968	0.99965	0.99962	0.99959
	100000	0.99968	0.99966	0.99964	0.99961	0.99958
	101325	0.99967	0.99966	0.99963	0.99961	0.99957
	105000	0.99966	0.99964	0.99962	0.99960	0.99956
	110000	0.99965	0.99963	0.99961	0.99958	0.99955
26.0	95000	0.99970	0.99968	0.99966	0.99963	0.99959
	100000	0.99969	0.99967	0.99964	0.99961	0.99958
	101325	0.99968	0.99966	0.99964	0.99961	0.99957
	105000	0.99967	0.99965	0.99963	0.99960	0.99956
	110000	0.99966	0.99964	0.99961	0.99959	0.99955

the saturation concentration of water vapor above the surface of the water; the effective saturation vapor pressure of water, e_s', is greater than the saturation vapor pressure of pure phase (i.e., water vapor without the admixture of air or any other substance), e_s.

ENHANCEMENT FACTOR, f

This "enhancement" of water vapor pressure is expressed by the enhancement factor, f, which is defined by

$$f = e_s'/e_s \tag{2.26}$$

A published experimentally derived value of f^{15} at 20°C and 100000 Pa is 1.00400. Therefore, the common practice of inferring e' from measured U and tabulated value of e_s introduces a significant error in e' if f has been ignored. The corresponding relative error in ρ at 20°C, 101325 Pa, and 50% relative humidity is about 1.7×10^{-5}.

f is a function of temperature and pressure. In the present work, Hyland's values of f^{15} have been fitted to a three-parameter equation in the pressure (P, Pa) and temperature (t, °C) ranges of interest in national standards laboratories. The resulting equation is

$$f = 1.00070 + 3.113 \times 10^{-8}\, P + 5.4 \times 10^{-7}\, t^2 \tag{2.27}$$

Over the temperature range 19.0 to 26.0°C and the pressure range 69,994 Pa (10.152 PSI) to 110,004 Pa (15.955 PSI), f ranges from 1.0031 to 1.0045. The maximum variation of f from a nominal value of 1.0042 is equal to 0.11% of the nominal value. The corresponding relative variation of air density is equal to 0.00040%, which is negligible.

SATURATION VAPOR PRESSURE OF WATER, e_s

The expression for e' is found by combining Equations 2.25 and 2.26 to be

$$e' = (U/100)fe_s \tag{2.28}$$

The systematic relative uncertainties in ρ due to the uncertainties assigned to f^{15} and $e_s{}^{18}$ are approximately $\pm 1 \times 10^{-6}$ and $\pm 2 \times 10^{-7}$, respectively.

The e_s data of Besley and Bottomley[19] in the temperature range 288.15K (15°C) to 298.04K (24.89°C) and calculated values for the remainder of the temperature range to 301.15K (28°C) have been fitted[1] to a two-parameter equation. The resulting equation is

$$e_s = 1.7526 \times 10^{11}\, e^{(-5315.56/T)} \tag{2.29}$$

where $e = 2.7182818...$ is the base of Naperian logarithms.

Values calculated using Equation 2.29 are sufficiently close to experimental and calculated values of e_s, within $\pm 0.1\%$, to be used in the present application. Calculated values of e_s expressed in mm Hg are tabulated in Table 2.3.

CARBON DIOXIDE ABUNDANCE, x_{CO_2}

In a previous section it has been stated that the CO_2 abundance in laboratory air is in general variable. A variation of 0.0001 in CO_2 mole fraction is equivalent to a relative variation of 4×10^{-5} in calculated air density. The CO_2 abundance should be known for optimum utilization of the air density calculation. Equation 2.18 can be used to adjust M_a for departures of CO_2 abundance from the reference level, 0.00033.

THE AIR DENSITY EQUATION

By combining Equations 2.11 and 2.28 and substituting M_w/M_a for ϵ, the air density equation developed by Jones[1] becomes

$$\rho = (PM_a/RTZ)[1 - (1 - M_w/M_a)(U/100)(fe_s/P)] \qquad (2.30)$$

By substituting the R value of Moldover et al., $8.314471 \text{ J mol}^{-1} \text{ K}^{-1}$, the value of 18.0152 for M_w, and the value of 28.963 for M_a, Equation 2.30 becomes

$$\rho = 0.000120272(PM_a/TZ)[1 - (1 - 18.0152/M_a)(U/100)(fe_s/P)] \quad (2.31)$$

where

$$M_a = 28.963 + 12.011 \ (x_{CO_2} - 0.00033) \qquad (2.32)$$

For $T = 293.15K$ (20°C), $P = 101325$ Pa, 50% relative humidity, and $M_a = 28.963 \text{ g mol}^{-1}$, the air density calculated using Equation 2.31 is 1.1992 kg/m^{-3} = 0.0011992 g/cm^3 = 1.1992 mg/cm^3.

For M_a of 28.964 g mol^{-1} ($x_{CO_2} = 0.00043$) and the same values of the other parameters above, $\rho = 1.1993$ kg/m^{-3} = 0.0011993 g/cm^{-3} = 1.1993 mg/cm^{-3}.

UNCERTAINTIES IN AIR DENSITY

We now return to the air density equation as expressed by Equation 2.30. The uncertainty propagation equation[20] as it applies to air density is

$$(SD_\rho)^2 = \Sigma_i[(\partial\rho/\partial Y_i)^2 \ (SD_i)^2] \qquad (2.33)$$

where SD_ρ is the estimate of standard deviation of ρ, the Y_i's are quantities on the right-hand side of Equation 2.30, and the SD_i's are the estimates of standard deviation of the quantities on the right-hand side of Equation 2.30.

TABLE 2.3
Saturation Vapor Pressure of Water
Calculated Using Equation 2.29

t (°C)	e$_s$ (mm Hg)	t (°C)	e$_s$ (mm Hg)
19.0	16.480	22.6	20.565
.1	16.583	.7	20.691
.2	16.686	.8	20.817
.3	16.790	.9	20.943
.4	16.895		
.5	17.000	23.0	21.071
.6	17.106	.1	21.199
.7	17.212	.2	21.327
.8	17.319	.3	21.457
.9	17.427	.4	21.587
		.5	21.718
20.0	17.535	.6	21.849
.1	17.644	.7	21.982
.2	17.753	.8	22.114
.3	17.863	.9	22.248
.4	17.974		
.5	18.085	24.0	22.383
.6	18.197	.1	22.518
.7	18.309	.2	22.653
.8	18.422	.3	22.790
.9	18.536	.4	22.927
		.5	23.065
21.0	18.650	.6	23.204
.1	18.765	.7	23.344
.2	18.880	.8	23.484
.3	18.996	.9	23.625
.4	19.113		
.5	19.231	25.0	23.767
.6	19.349	.1	23.909
.7	19.467	.2	24.052
.8	19.587	.3	24.196
.9	19.707	.4	24.341
		.5	24.487
22.0	19.827	.6	24.633
.1	19.949	.7	24.780
.2	20.071	.8	24.928
.3	20.193	.9	25.077
.4	20.317		
.5	20.441	26.0	25.226

UNCERTAINTIES IN QUANTITIES OTHER THAN P, T, U, AND X_{CO_2}

We first concentrate on the uncertainties in the quantities other than the measurements of P, T, U, and x_{CO_2} to find the minimum uncertainty in ρ if P, T, U, and x_{CO_2} were measured perfectly. This minimum uncertainty can then be considered the limitation on the determination of ρ using the air density equation. We shall refer to these quantities loosely as nonenvironmental quantities.

Partial Derivatives, $(\partial\rho/\partial Y_i)$, for the Nonenvironmental Quantities

The partial derivatives of ρ with respect to the nonenvironmental quantities are now taken and evaluated using the following values of the parameters:

P	101325 Pa
M_a	28.963 g/mol
R	8314.471 J/kmol/K
T	293.15K
Z	0.99960
M_w	18.0152 g/mol
U	50%
f	1.0040
e_s	2338.80 Pa
x_{CO_2}	0.00033

The value of ρ calculated using the quantities above is 1.1992 kg/m^3. The partial derivatives for the nonenvironmental quantities are

$$(\partial\rho/\partial M_a) = (1/RTZ)[P - (Ufe_s/100)] = 4.1106 \times 10^{-2}$$

$$(\partial\rho/\partial R) = (1/R^2TZ)[(M_a - M_w)(Ufe_s/100) - PM_a] = -1.4423 \times 10^{-4}$$

$$(\partial\rho/\partial Z) = (1/RTZ^2)[(M_a - M_w)(Ufe_s/100) - PM_a] = -1.1997$$

$$(\partial\rho/\partial f) = [Ue_s/(100\ RTZ)](M_w - M_a) = -5.2546 \times 10^{-3}$$

$$(\partial\rho/\partial e_s) = [Uf/(100\ RTZ)](M_w - M_a) = -2.2557 \times 10^{-6}$$

$$(\partial\rho/\partial M_w) = (1/RTZ)(Ufe_s/100) = 4.8189 \times 10^{-4}$$

For convenience, units have not been given for the partial derivatives.

UNCERTAINTIES IN THE NONENVIRONMENTAL QUANTITIES, (SD_i)

The estimates of the standard deviations for the nonenvironmental quantities are given below:

M_a 1×10^{-3} g/mol
R 1.4×10^{-2} J/kmol/K
Z 5.7×10^{-6}
f 7×10^{-5}
e_s 3.9×10^{-2} Pa
M_w 2×10^{-4} g/mol

Products of the Partial Derivatives and the Estimates of Standard Deviation, $(\partial\rho/\partial Y_i)\cdot(SD_i)$, for the Nonenvironmental Quantities

The product of the partial derivative and the estimate of standard deviation for each of the nonenvironmental quantities is given below:

M_a 4×10^{-5} kg/m^3
R -2.0×10^{-6} kg/m^3
Z -6.8×10^{-6} kg/m^3
f -4×10^{-7} kg/m^3
e_s -8.8×10^{-8} kg/m^3
M_w 1×10^{-7} kg/m^3

The above products were squared and summed as indicated in Equation 2.33 and the square root was taken. The resulting estimate of the standard deviation of the air density, ρ, is 4×10^{-5} kg/m^3. The estimate of standard deviation of ρ is dominated by the uncertainty in the apparent molecular weight of dry air, M_a.

Therefore, the minimum uncertainty in the calculation of air density using the air density equation, that which can be considered to be intrinsic to the equation, is an estimate of standard deviation of 4×10^{-5} kg/m^3 or 34 parts per million in the present example. Note that this does not include uncertainties in the measurements of pressure, temperature, humidity, and mole fraction of CO_2.

UNCERTAINTIES IN THE ENVIRONMENTAL QUANTITIES
Partial Derivatives, $\partial\rho/\partial Y_i$, for the Environmental Quantities

The partial derivatives for the environmental quantities, P, T, U, and x_{CO_2} are

$$(\partial\rho/\partial P) = (M_a/RTZ) = 1.1888 \times 10^{-5}$$

$$(\partial\rho/\partial T) = (1/RT^2Z)[-PM_a + (M_a - M_w)(Ufe_s/100)]$$

$$= -4.0908 \times 10^{-3}$$

$$(\partial\rho/\partial U) = (1/RTZ)(fe_s/100)(M_w - M_a) = -1.0551 \times 10^{-4}$$

Again, for convenience, the units for the partial derivatives have not been given.

For atmospheres in which the mole fraction of CO_2, x_{CO_2}, is other than the reference value of 0.00033,

$$M_a = 28.963 + 12.011 \ (x_{CO_2} - 0.00033) \tag{2.34}$$

and

$$(\partial M_a / \partial x_{CO_2}) = 12.011$$

Then,

$$(\partial \rho / \partial x_{CO_2}) = (\partial \rho / \partial M_a) \ (\partial M_a / \partial x_{CO_2}) = (4.1106 \times 10^{-2}) \ (12.011)$$

$$= 0.49372$$

UNCERTAINTIES IN THE ENVIRONMENTAL QUANTITIES, (SD$_i$)

The estimates of the standard deviations of each of the four environmental parameters are

P = 20.5 Pa
T = 0.05 K
U = 0.51% RH
x_{CO_2} = 0.00005

Products of the Partial Derivatives and the Estimates of Standard Deviation, $(\partial \rho / \partial Y_i) \cdot (SD_i)$, for the Environmental Quantities

The product of the partial derivative and the estimate of standard deviation for each of the environmental parameters is

P = 2.4×10^{-4} kg/m^3
T = -2.0×10^{-4} kg/m^3
U = -5.4×10^{-5} kg/m^3
x_{CO_2} = 2.5×10^{-5} kg/m^3

The above products were squared and summed, and the square root was taken. The resulting estimate of the standard deviation of the air density, ρ, is 3×10^{-4} kg/m^3.

The largest uncertainty in ρ is that due to pressure measurement; the next largest is that due to temperature measurement. The estimates of uncertainty are based on state-of-the-art measurements of the variables.

The overall uncertainty contributed by all of the quantities in the air density equation is 3×10^{-4} kg/m^3.

USE OF CONSTANT VALUES OF F, Z, AND M_a IN THE AIR DENSITY EQUATION

By considering the expected variations in pressure, temperature, and relative humidity in the laboratory, it might be possible to use constant values of f, Z, and M_a.

For example, in the Mass Laboratory of NIST, constant values of f (1.0042), Z (0.99966), and M_a (28.964) are considered to be adequate. With these values of f, Z, and M_a, the resulting equation for calculating air density is, for P in pascals and absolute temperature T = 273.15 + t (°C),

$$\rho_a = (0.0034847/T)(P - 0.0037960\ Ue_s) \qquad (2.35)$$

For P in PSI and t in °C,

$$\rho_a = [24.026/(t + 273.15)](P - 0.0037960\ Ue_s) \qquad (2.36)$$

For P = 101325 Pa = 14.69595 PSI, T = 293.15 (t = 20°C), U=50, and e_s=2338.80Pa, ρ_a = 1.1992 kg/m^{-3} = 0.0011992 g/cm^{-3} = 1.1992 mg/cm^{-3} for both Equations 2.35 and 2.36.

CIPM-81 AIR DENSITY EQUATION

The Comite International des Poids et Mesures recommended the use of what has become known as the CIPM-81 air density equation.[21] The CIPM-81 equation was based on the Jones equation[1] and the differences between the two equations are not major. The modified Jones equation, Equation 2.31, is recommended here for use for the calculation of the density of moist air.

DENSITIES OF SOME OTHER GASES

The general equation for the density of a gas other than air is

$$\rho = PM/RTZ \qquad (2.37)$$

where M is the molecular or atomic weight of the gas and the other symbols are as above.

DENSITY OF ARGON

The atomic weight of argon is 39.948 g mol^{-1} and R = 8314.471 J kmol^{-1} K^{-1}. Values of Z for argon are given by Perry and Chilton;[22] for example, at 300K (26.85°C) and 1 atmosphere (101,325 Pa, 14.69595 PSI) the value of Z given is 0.9994. The calculated value of ρ using Equation 2.37 is 1.624 kg/m^3 = 0.001624 g/cm^3 = 1.624 mg/cm^3.

DENSITY OF CARBON DIOXIDE
The molecular weight of carbon dioxide is 44.0098 g mol^{-1}. Values of Z for carbon dioxide are given in Reference 22; for example, at 0°C (273.15K) and 100,000 Pa (14.50378 PSI) the value of Z given is 0.9933. The calculated value of ρ using Equation 2.37 is 1.951 kg/m^3 = 0.001951 g/cm^3 = 1.951 mg/cm^3.

DENSITY OF HYDROGEN
The molecular weight of hydrogen is 2.0158 g mol^{-1}. Values of Z for hydrogen are given in Reference 22; for example, at 300K (26.85°C) and 1 atmosphere (101,325 Pa, 14.69595 PSI) the value of Z given is 1.0006. The calculated value of ρ using Equation 2.37 is 0.081837 kg/m^3 = 0.000081837 g/cm^3 = 0.081837 mg/cm^3.

DENSITY OF CARBON MONOXIDE
The molecular weight of carbon monoxide is 28.0104 g mol^{-1}. Values of Z for carbon monoxide are given in Reference 22; for example, at 300K (26.85°C) and 1 atmosphere (101,325 Pa, 14.69595 PSI) the value of Z given is 0.9997. The calculated value of ρ using Equation 2.37 is 1.138 kg/m^3 = 0.001138 g/cm^3 = 1.138 mg/cm^3.

DENSITY OF METHANE
The molecular weight of methane is 16.0426 g mol^{-1}. Values of Z for methane are given in Reference 22; for example, at 300K (26.85°C) and 100,000 Pa (1 bar, 14.50378 PSI) the value of Z given is 0.9982. The calculated value of ρ using Equation 2.37 is 0.6443 kg/m^3 = 0.0006443 g/cm^3 = 0.6443 mg/cm^3.

DENSITY OF NITROGEN
The molecular weight of nitrogen is 28.0134 g mol^{-1}. Values of Z for nitrogen are given in Reference 22; for example, at 300K (26.85°C) and 100,000 Pa (1 bar, 14.50378 PSI), the value of Z given is 0.9998. The calculated value of ρ using Equation 2.37 is 1.123 kg/m^3 = 0.001123 g/cm^3 = 1.123 mg/cm^3.

DENSITY OF OXYGEN
The molecular weight of oxygen is 31.9988 g mol^{-1}. Values of Z for oxygen are given in Reference 22; for example, at 300K (26.85°C) and 100,000 Pa (1 bar, 14.50378 PSI), the value of Z given is 0.9994. The calculated value of ρ using Equation 2.37 is 1.284 kg/m^3 = 0.001284 g/cm^3 = 1.284 mg/cm^3.

EXERCISES

1. If air were an ideal gas (i.e., Z = 1), what would its density be at 20°C and 14.69595 PSI?
2. If nitrogen were an ideal gas, what would its density be at 20°C and 14.69595 PSI?

3. If oxygen and nitrogen were ideal gases, what would be the ratio of their densities at 20°C and 14.69595 PSI? What would be the ratio of their molecular weights? How are these two numbers related?

4. If all of the eight gases above (including dry air) were ideal gases, rank their densities from lowest to highest. Rank their molecular weights from lowest to highest. How are these rankings related?

5. If density, P, R, T, and M were known for a gas, how would one infer the compressibility of the gas?

6. If the relative uncertainties for each of the above five quantities (not Z) were known, what would the relative uncertainty of Z be?

7. If the density of a gas were determined independently, could one determine Z?

8. If the other quantities in a real gas equation were known, could one infer the value of R, the universal gas constant? What would be the limiting uncertainty in the inference of R?

9. All other things being equal, in which of the above gases should a weighing be made to minimize the effect of not making a buoyancy correction?

REFERENCES

1. **Jones, F. E.,** An air density equation and the transfer of the mass unit, *J. Res. Nat. Bur. Stand.*, 83, 419, 1978.

2. **List, R. J.,** *Smithsonian Meteorological Tables*, 6th rev. ed., Smithsonian Institution, Washington, DC, 1951, 347.

3. **Cohen, E. R. and Taylor, B. N.,** The 1993 least-squares adjustment of the fundamental constants, *J. Phys. Chem. Ref. Data*, 2, 663, 1973.

4. **Quinn, T. J., Colclough, A. R., and Chandler, T. R. D.,** A new determination of the gas constant by an acoustical method, *Phil. Trans. R. Soc. London*, A283, 367, 1976.

5. **Gammon, B. E.,** The velocity of sound with derived state properties in helium at −175 to 159°C with pressure to 150 Atm, *J. Chem. Phys.*, 64, 2556, 1976.

6. **Gammon, B. E.,** private communication.

7. **Rowlinson, J. S. and Tildesley, D. J.,** Determination of the gas constant from the speed of sound, *Proc. R. Soc. London*, A358, 281, 1977.

8. **Moldover, M. R., Trusler, J. P. M., Edwards, T. J., Mehl, J. B., and Davis, R. S.,** Measurement of the universal gas constant R using a spherical acoustic resonator, *Phys. Rev. Lett.*, 60, 249, 1988.

9. **Moldover, M. R., Trusler, J. P. M., Edwards, T. J., Mehl, J. B., and Davis, R. S.,** Measurement of the universal gas constant R using a spherical acoustic resonator, *J. Res. Natl. Bur. Stand.*, 93, 85, 1988.

10. International Union of Pure and Applied Chemistry, Inorganic Chemistry Division, Commission on Atomic Weights, *Pure and Appl. Chem.*, 47, 75, 1976.

11. **Machta, L. and Hughes, E.,** Atmospheric oxygen in 1967 to 1970, *Science*, 168, 1582, 1970.

12. **Gluekauf, E.,** *Compendium of Meteorology*, Malone, T. F., Ed., The American Meteorological Society, Boston, 1951, 4.

13. **Hughes, E.,** Unpublished paper, presented only at the American Chemical Society Fourth Mid-Atlantic Meeting, Washington, DC, February 1969.
14. *U. S. Standard Atmosphere, 1976*, U. S. Government Printing Office, Washington, DC, 1976, 3 and 33.
15. **Hyland, R. W.,** A correlation for the second interaction virial coefficients and the enhancement factors for moist air, *J. Res. Natl. Bur. Stand.*, 79A, 551, 1975.
16. **List, R. J.,** *Smithsonian Meteorological Tables*, 6th rev. ed., Smithsonian Institution, Washington, DC, 1951, 352.
17. **Harrison, L. P.,** *Humidity and Moisture*, Vol. 3, Wexler, A. and Wildhack, W. A., Eds., Reinhold, New York, 1964, 51.
18. **Wexler, A. and Greenspan, L.,** Vapor pressure equation for water vapor in the range 0 to 100°C, *J. Res. Natl. Bur. Stand.*, 75A, 213, 1971.
19. **Besley, L. and Bottomley, G. A.,** Vapour pressure of normal and heavy water from 273.15 to 298.15 K, *J. Chem. Thermodyn.*, 5, 397, 1973.
20. **Eisenhart, C. J.,** Realistic evaluation of the precision and accuracy of instrument calibration systems, *J. Res. Natl. Bur. Stand.*, 67C, 161, 1963.
21. **Giacomo, P.,** Equation for the determination of the density of moist air (1981), *Metrologia*, 18, 33, 1982.
22. **Perry, R. H. and Chilton, C. H.,** *Chemical Engineers' Handbook*, McGraw-Hill, New York, 1973.

Chapter 3

DENSITY OF WATER

INTRODUCTION

The density of water is used as a reference in various areas of flow measurement. For example, water is used as the calibrating fluid in the calibration (gravimetric determination) of the volume of volumetric standards. The volume is calculated from the mass and the density of the water. The formulation of Wagenbreth and Blanke[1] is used in many quarters to calculate the density of water.

In this chapter, a formulation[2] based on the work of Kell[3] is presented. In flow measurement work, except in the rare case in which freshly distilled water is used, the water density of interest is the density of air-saturated water. In the present formulation, equations were developed for the density of air-saturated water as a function of temperature on the 1990 International Temperature Scale (ITS-90)[4] and for the isothermal compressibility of water on ITS-90. The temperature range for the formulation is 5 to 40°C.

FORMULATIONS OF KELL

DENSITY OF AIR-FREE WATER

Kell[3] published a new formulation for the density of *air-free* water valid from 0 to 150°C at a pressure of 101.325 kPa (1 atmosphere) "that is in improved agreement with most data sets:"[3]

$$\rho \ (kg \ m^{-3}) = (999.83952 + 16.945176 \ t - 7.9870401 \times 10^{-3} \ t^2$$

$$- 46.170461 \times 10^{-6} \ t^3 + 105.56302 \times 10^{-9} \ t^4$$

$$- 280.54253 \times 10^{-12} \ t^5)/(1 + 16.897850 \times 10^{-3} \ t) \quad (3.1)$$

where ρ is the density of air-free water and t is temperature in °C on the 1968 International Practical Temperature Scale (IPTS-68).

Isothermal Compressibility of Air-Free Water

An equation for calculation of the isothermal compressibility, κ_T, for air-free water also was developed by Kell.[3] The equation can be expressed, in the temperature range of 0 to 100°C on IPTS-68, as

$$\kappa_T = (50.88496 \times 10^{-8} + 6.163813 \times 10^{-9} \ t + 1.459187 \times 10^{-11} \ t^2$$

$$+ 20.08438 \times 10^{-14} \ t^3 - 58.47727 \times 10^{-17} \ t^4$$

$$+ 410.4110 \times 10^{-20} \ t^5)/(1 + 19.67348 \times 10^{-3} \ t) \quad (3.2)$$

where κ_T is the isothermal compressibility of air-free water in $(kPa)^{-1}$.

NEW FORMULATIONS

DENSITY OF AIR-FREE WATER

In more recent work,[2] values of ρ calculated by Kell[3] were fitted to an equation quartic in temperature at a pressure of 101.325 kPa over the temperature range 5 to 40°C on ITS-90:

$$\rho \; (\text{kg m}^{-3}) = 999.85308 + 6.32693 \times 10^{-2} \, t - 8.523829 \times 10^{-3} \, t^2$$

$$+ \; 6.943248 \times 10^{-5} \, t^3 - 3.821216 \times 10^{-7} \, t^4 \qquad (3.3)$$

Equation 3.3 applies to *air-free* water. Unlike Equation 3.1, a term in t^5 is not necessary, due at least in part to the fact that the 0 to 4°C region, in which ρ increases with increasing temperature, has been excluded.

Values of density of air-free water were calculated using Equation 3.3 for temperatures (ITS-90, t_{90}) of 4.999 to 39.990°C. These values were compared with corresponding Kell[3] values. The estimate of the standard deviation (SD) for the differences between Equations 3.3 and 3.1 was found to be 0.00034 kg/m³. The ratio of SD to the mean value of water density was 0.34 ppm (parts per million) which is negligible.

CONVERSION OF IPTS-68 TO ITS-90

A very simple equation[2] relating ITS-90 temperature,[4] t_{90}, to IPTS-68 temperature, t_{68}, has been developed and used to generate values of t_{90} for the development of Equation 3.3 and following equations.[2]

In the temperature range 0 to 40°C, the conversion equation is

$$t_{90} = 0.0002 + 0.99975 \, t_{68} \qquad (3.4a)$$

In the temperature range 0 to 100°C, the conversion equation is

$$t_{90} = 0.0005 + 0.9997333 \, t_{68} \qquad (3.4b)$$

CHANGE IN DENSITY OF WATER WITH AIR SATURATION

For 80 points in the temperature range 4 to 20°C, Bignell[5] measured the change in the density of water with air saturation and fitted the points to develop the equation

$$\Delta\rho = -0.004612 + 0.000106 \, t \qquad (3.5)$$

where $\Delta\rho$ is the change in the density of water with air density, in kg m⁻³. Bignell concluded that "there is probably not much need to extend the work to higher temperatures because the effect diminishes and the accuracy of density metrology [in 1983] at these temperatures would not warrant a more accurately known correction."

DENSITY OF AIR-SATURATED WATER

To produce an equation to be used to calculate the density, ρ_{as}, of *air-saturated* water in the temperature range 5 to 40°C on ITS-90, Equation 3.5 was added to Equation 3.3.[2] The resulting equation is

$$\rho_{as} = 999.84847 + 6.337563 \times 10^{-2}\, t - 8.523829 \times 10^{-3}\, t^2$$

$$+ 6.943248 \times 10^{-5}\, t^3 - 3.821216 \times 10^{-7}\, t^4 \tag{3.6}$$

At 20°C, the uncertainty in the density of air-saturated water for an uncertainty in temperature of 1°C is approximately 210 ppm or 0.21 kg m^{-3}.

ISOTHERMAL COMPRESSIBILITY OF WATER

Using the data for isothermal compressibility used by Kell,[3] an equation to be used to calculate the isothermal compressibility of water for the temperature range 5 to 40°C on ITS-90 was developed:[2]

$$\kappa_T = 50.83101 \times 10^{-8} - 3.68293 \times 10^{-9}\, t + 7.263725 \times 10^{-11}\, t^2$$

$$- 6.597702 \times 10^{-13}\, t^3 + 2.87767 \times 10^{-15}\, t^4 \tag{3.7}$$

where κ_T is in (kPa)$^{-1}$.

The estimate of the SD of the difference between calculated values of κ_T and data values is 2.1×10^{-11} (kPa)$^{-1}$. The ratio of SD to the midrange value of κ_T is 4.6×10^{-5}, which is negligible for present purposes. It is not necessary to make a correction to κ_T for air saturation.

At 20°C and a pressure of 1 atmosphere, the value of the isothermal compressibility of water is approximately 46.5 ppm/atmosphere. A correction for isothermal compressibility calculated using Equation 3.7 should be made to density of water calculations made for locations where the atmospheric pressure is significantly different from 1 atmosphere (101.325 kPa). For example, at Boulder, CO, the correction for isothermal compressibility of water is approximately −8 ppm at 20°C.

COMPRESSIBILITY-CORRECTED
DENSITY OF WATER EQUATION

For ρ_{as} calculated using Equation 3.6 and κ_T calculated using Equation 3.7, the expression for the density of air-saturated water corrected for isothermal compressibility, ρ_{asc}, is

$$\rho_{asc} = \rho_{as}\, [1 + \kappa_T\, (P - 101.325)] \tag{3.8}$$

where P is ambient atmospheric pressure in kPa.

TABLES

Values of the density of air-free water calculated using Equation 3.3 are tabulated in Table 3.1. The values are given, in this and the other tables, in units g/cm^3. Values of the density of air-saturated water calculated using Equation 3.6 are tabulated in Table 3.2. For comparison purposes, values of the density of air-free water given by Wagenbreth and Blanke[1] are tabulated in Table 3.3.

SUMMARY

Equation 3.6 can be used to calculate the density of *air-saturated* water in the temperature range 5 to 40°C at a pressure of 1 atmosphere (101.325 Pa). The units for water density are kg m^{-3} in this development, although the units in the tables are g cm^{-3}.

Equation 3.8, which includes Equation 3.7, can be used to calculate the density of air-saturated water in the temperature range 5 to 40°C at ambient atmosphere pressures of P kPa.

Equation 3.3 can be used to calculate the density of *air-free* water in the temperature range 5 to 40°C at a pressure of 1 atmosphere. An equation of the form of Equation 3.8 can be used to calculate the density of air-free water at other ambient pressures. Equation 3.7 would be applied unchanged.

REFERENCES

1. **Wagenbreth, H. and Blanke, W.,** Die Dichte des Wassers im Internationalen Einheitensystem und im der Internationalen Praktischen Temperaturkala von 1968, *PTB Mitt*, 81, 412, 1971.
2. **Jones, F. E. and Harris, G. L.,** The air density equation and the transfer of the mass unit, *J. Res. Natl. Inst. Stand. Technol.*, 97, 335, 1992.
3. **Kell, G. S.,** Density, thermal expansivity, and compressibility of liquid water from 0° to 150°C: correlations and tables for atmospheric pressure and saturation reviewed and expressed on 1968 Temperature Scale, *J. Chem. Eng. Data*, 20, 97, 1975.
4. **Mangum, B. W. and Furukawa, G. T.,** Guidelines for realizing the International Temperature Scale of 1990 (ITS-90), *Natl. Inst. Stand. Technol. Technical Note 1265*, 1990.
5. **Bignell, N.,** The effect of dissolved air on the density of water, *Metrologia*, 19, 57, 1983.

TABLE 3.1
Density of Air-Free Water (g/cm³), Equation (3.3)

°C	0.0	0.1	0.2	0.3	0.4	0.5	0.6	0.7	0.8	0.9
5	0.999965	0.999963	0.999961	0.999959	0.999957	0.999954	0.999952	0.999949	0.999946	0.99994
6	0.999940	0.999937	0.999934	0.999930	0.999926	0.999923	0.999919	0.999914	0.999910	0.99990
7	0.999901	0.999896	0.999892	0.999887	0.999881	0.999876	0.999871	0.999865	0.999860	0.99985
8	0.999848	0.999842	0.999835	0.999829	0.999822	0.999816	0.999809	0.999802	0.999795	0.99978
9	0.999780	0.999773	0.999765	0.999757	0.999749	0.999741	0.999733	0.999725	0.999716	0.99970
10	0.999699	0.999690	0.999681	0.999672	0.999663	0.999653	0.999644	0.999634	0.999624	0.99961
11	0.999604	0.999594	0.999584	0.999574	0.999563	0.999552	0.999541	0.999531	0.999519	0.99950
12	0.999497	0.999485	0.999474	0.999462	0.999450	0.999438	0.999426	0.999414	0.999402	0.99938
13	0.999377	0.999364	0.999351	0.999338	0.999325	0.999312	0.999299	0.999285	0.999272	0.99925
14	0.999244	0.999230	0.999216	0.999202	0.999188	0.999173	0.999159	0.999144	0.999129	0.99911
15	0.999099	0.999084	0.999069	0.999053	0.999038	0.999022	0.999007	0.998991	0.998975	0.99895
16	0.998943	0.998926	0.998910	0.998893	0.998877	0.998860	0.998843	0.998826	0.998809	0.99879
17	0.998774	0.998757	0.998739	0.998722	0.998704	0.998686	0.998668	0.998650	0.998632	0.99861
18	0.998595	0.998576	0.998558	0.998539	0.998520	0.998501	0.998482	0.998463	0.998444	0.99842
19	0.998405	0.998385	0.998365	0.998345	0.998325	0.998305	0.998285	0.998265	0.998244	0.99822
20	0.998203	0.998183	0.998162	0.998141	0.998120	0.998099	0.998077	0.998056	0.998035	0.99801
21	0.997991	0.997970	0.997948	0.997926	0.997904	0.997882	0.997859	0.997837	0.997815	0.99779
22	0.997769	0.997746	0.997724	0.997701	0.997678	0.997654	0.997631	0.997608	0.997584	0.99756
23	0.997537	0.997513	0.997489	0.997465	0.997441	0.997417	0.997393	0.997369	0.997344	0.99732
24	0.997295	0.997270	0.997245	0.997220	0.997195	0.997170	0.997145	0.997120	0.997094	0.99706

TABLE 3.1 (CONTINUED)
Density of Air-Free Water (g/cm³), Equation (3.3)

°C	0.0	0.1	0.2	0.3	0.4	0.5	0.6	0.7	0.8	0.9
25	0.997043	0.997017	0.996992	0.996966	0.996940	0.996914	0.996887	0.996861	0.996835	0.99680
26	0.996782	0.996755	0.996728	0.996701	0.996675	0.996648	0.996620	0.996593	0.996566	0.99653
27	0.996511	0.996483	0.996456	0.996428	0.996400	0.996372	0.996344	0.996316	0.996288	0.99626
28	0.996231	0.996203	0.996174	0.996146	0.996117	0.996088	0.996059	0.996030	0.996001	0.99597
29	0.995942	0.995913	0.995884	0.995854	0.995824	0.995795	0.995765	0.995735	0.995705	0.99567
30	0.995645	0.995615	0.995584	0.995554	0.995523	0.995493	0.995462	0.995431	0.995401	0.99537
31	0.995339	0.995307	0.995276	0.995245	0.995214	0.995182	0.995151	0.995119	0.995087	0.99505
32	0.995024	0.994992	0.994960	0.994928	0.994895	0.994863	0.994831	0.994798	0.994766	0.99473
33	0.994701	0.994668	0.994635	0.994602	0.994569	0.994536	0.994503	0.994469	0.994436	0.99440
34	0.994369	0.994335	0.994302	0.994268	0.994234	0.994200	0.994166	0.994132	0.994098	0.99406
35	0.994029	0.993995	0.993960	0.993926	0.993891	0.993856	0.993822	0.993787	0.993752	0.99371
36	0.993682	0.993646	0.993611	0.993576	0.993540	0.993505	0.993469	0.993433	0.993398	0.99336
37	0.993326	0.993290	0.993254	0.993217	0.993181	0.993145	0.993108	0.993072	0.993035	0.99299
38	0.992962	0.992925	0.992888	0.992851	0.992814	0.992777	0.992740	0.992703	0.992665	0.99262
39	0.992590	0.992553	0.992515	0.992478	0.992440	0.992402	0.992364	0.992326	0.992288	0.99224

TABLE 3.2

Density of Air-Saturated Water (g/cm³), Equation (3.6)

°C	0.0	0.1	0.2	0.3	0.4	0.5	0.6	0.7	0.8	0.9
5	0.999961	0.999959	0.999957	0.999955	0.999953	0.999950	0.999948	0.999945	0.999942	0.99993
6	0.999936	0.999933	0.999930	0.999926	0.999922	0.999919	0.999915	0.999911	0.999906	0.99990
7	0.999897	0.999893	0.999888	0.999883	0.999878	0.999872	0.999867	0.999861	0.999856	0.99985
8	0.999844	0.999838	0.999832	0.999825	0.999819	0.999812	0.999805	0.999798	0.999791	0.99978
9	0.999777	0.999769	0.999761	0.999754	0.999746	0.999738	0.999730	0.999721	0.999713	0.99970
10	0.999695	0.999687	0.999678	0.999669	0.999659	0.999650	0.999640	0.999631	0.999621	0.99961
11	0.999601	0.999591	0.999581	0.999570	0.999560	0.999549	0.999538	0.999527	0.999516	0.99950
12	0.999494	0.999482	0.999471	0.999459	0.999447	0.999435	0.999423	0.999411	0.999398	0.99938
13	0.999373	0.999361	0.999348	0.999335	0.999322	0.999309	0.999295	0.999282	0.999268	0.99925
14	0.999241	0.999227	0.999213	0.999199	0.999184	0.999170	0.999156	0.999141	0.999126	0.99911
15	0.999096	0.999081	0.999066	0.999051	0.999035	0.999019	0.999004	0.998988	0.998972	0.99895
16	0.998940	0.998923	0.998907	0.998891	0.998874	0.998857	0.998840	0.998823	0.998806	0.99878
17	0.998772	0.998754	0.998737	0.998719	0.998701	0.998683	0.998665	0.998647	0.998629	0.99861
18	0.998592	0.998574	0.998555	0.998536	0.998517	0.998499	0.998479	0.998460	0.998441	0.99842
19	0.998402	0.998382	0.998363	0.998343	0.998323	0.998303	0.998283	0.998262	0.998242	0.99822
20	0.998201	0.998180	0.998159	0.998138	0.998117	0.998096	0.998075	0.998054	0.998032	0.99801
21	0.997989	0.997967	0.997945	0.997924	0.997901	0.997879	0.997857	0.997835	0.997812	0.99779
22	0.997767	0.997744	0.997721	0.997698	0.997675	0.997652	0.997629	0.997606	0.997582	0.99755
23	0.997535	0.997511	0.997487	0.997463	0.997439	0.997415	0.997391	0.997366	0.997342	0.99731
24	0.997293	0.997268	0.997243	0.997218	0.997193	0.997168	0.997143	0.997118	0.997092	0.99706

TABLE 3.2 (CONTINUED)
Density of Air-Saturated Water (g/cm³), Equation (3.6)

°C	0.0	0.1	0.2	0.3	0.4	0.5	0.6	0.7	0.8	0.9
25	0.997041	0.997015	0.996990	0.996964	0.996938	0.996912	0.996885	0.996859	0.996833	0.99680
26	0.996780	0.996753	0.996726	0.996700	0.996673	0.996646	0.996619	0.996591	0.996564	0.99653
27	0.996509	0.996482	0.996454	0.996426	0.996399	0.996371	0.996343	0.996314	0.996286	0.99625
28	0.996230	0.996201	0.996173	0.996144	0.996115	0.996086	0.996057	0.996028	0.995999	0.99597
29	0.995941	0.995912	0.995882	0.995853	0.995823	0.995793	0.995764	0.995734	0.995704	0.99567
30	0.995643	0.995613	0.995583	0.995553	0.995522	0.995491	0.995461	0.995430	0.995399	0.99536
31	0.995337	0.995306	0.995275	0.995244	0.995212	0.995181	0.995149	0.995118	0.995086	0.99505
32	0.995023	0.994991	0.994959	0.994927	0.994894	0.994862	0.994830	0.994797	0.994765	0.99473
33	0.994699	0.994667	0.994634	0.994601	0.994568	0.994535	0.994502	0.994468	0.994435	0.99440
34	0.994368	0.994334	0.994301	0.994267	0.994233	0.994199	0.994165	0.994131	0.994097	0.99406
35	0.994028	0.993994	0.993960	0.993925	0.993890	0.993856	0.993821	0.993786	0.993751	0.99371
36	0.993681	0.993646	0.993610	0.993575	0.993539	0.993504	0.993468	0.993433	0.993397	0.99336
37	0.993325	0.993289	0.993253	0.993217	0.993181	0.993144	0.993108	0.993071	0.993035	0.99299
38	0.992962	0.992925	0.992888	0.992851	0.992814	0.992777	0.992740	0.992702	0.992665	0.99262
39	0.992590	0.992553	0.992515	0.992477	0.992439	0.992401	0.992363	0.992325	0.992287	0.99224

TABLE 3.3
Density of Air-Free Water (g/cm³), Wagenbreth and Blanke

°C	0.0	0.1	0.2	0.3	0.4	0.5	0.6	0.7	0.8	0.9
5	0.999964	0.999962	0.999960	0.999958	0.999956	0.999954	0.999951	0.999949	0.999946	0.99994
6	0.999940	0.999937	0.999933	0.999930	0.999926	0.999922	0.999918	0.999914	0.999910	0.99990
7	0.999901	0.999896	0.999892	0.999887	0.999881	0.999876	0.999871	0.999865	0.999860	0.99985
8	0.999848	0.999842	0.999835	0.999829	0.999822	0.999816	0.999809	0.999802	0.999795	0.99978
9	0.999780	0.999773	0.999765	0.999757	0.999749	0.999741	0.999733	0.999725	0.999716	0.99970
10	0.999699	0.999690	0.999681	0.999672	0.999662	0.999653	0.999643	0.999634	0.999624	0.99961
11	0.999604	0.999594	0.999583	0.999573	0.999562	0.999552	0.999541	0.999530	0.999519	0.99950
12	0.999496	0.999485	0.999473	0.999461	0.999449	0.999437	0.999425	0.999413	0.999401	0.99938
13	0.999376	0.999363	0.999350	0.999337	0.999324	0.999311	0.999297	0.999284	0.999270	0.99925
14	0.999243	0.999229	0.999215	0.999200	0.999186	0.999172	0.999157	0.999142	0.999128	0.99911
15	0.999098	0.999083	0.999067	0.999052	0.999036	0.999021	0.999005	0.998989	0.998973	0.99895
16	0.998941	0.998925	0.998908	0.998892	0.998875	0.998858	0.998841	0.998824	0.998807	0.99879
17	0.998773	0.998755	0.998738	0.998720	0.998702	0.998684	0.998666	0.998648	0.998630	0.99861
18	0.998593	0.998575	0.998556	0.998537	0.998519	0.998500	0.998480	0.998461	0.998442	0.99842
19	0.998403	0.998383	0.998364	0.998344	0.998324	0.998304	0.998284	0.998263	0.998243	0.99822
20	0.998202	0.998181	0.998160	0.998139	0.998118	0.998097	0.998076	0.998055	0.998033	0.99801
21	0.997990	0.997968	0.997947	0.997925	0.997903	0.997881	0.997858	0.997836	0.997814	0.99779
22	0.997768	0.997746	0.997723	0.997700	0.997677	0.997654	0.997630	0.997607	0.997584	0.99756
23	0.997536	0.997513	0.997489	0.997465	0.997441	0.997417	0.997392	0.997368	0.997344	0.99731
24	0.997294	0.997270	0.997245	0.997220	0.997195	0.997170	0.997145	0.997119	0.997094	0.99706

TABLE 3.3 (CONTINUED)
Density of Air-Free Water (g/cm³), Wagenbreth and Blanke

°C	0.0	0.1	0.2	0.3	0.4	0.5	0.6	0.7	0.8	0.9
25	0.997043	0.997017	0.996991	0.996966	0.996940	0.996913	0.996887	0.996861	0.996835	0.99680
26	0.996782	0.996755	0.996728	0.996702	0.996675	0.996648	0.996621	0.996593	0.996566	0.99653
27	0.996511	0.996484	0.996456	0.996428	0.996401	0.996373	0.996345	0.996316	0.996288	0.99626
28	0.996232	0.996203	0.996175	0.996146	0.996117	0.996088	0.996060	0.996031	0.996001	0.99597
29	0.995943	0.995914	0.995884	0.995855	0.995825	0.995795	0.995765	0.995736	0.995706	0.99567
30	0.995645	0.995615	0.995585	0.995554	0.995524	0.995493	0.995463	0.995432	0.995401	0.99537
31	0.995339	0.995308	0.995277	0.995246	0.995214	0.995183	0.995151	0.995120	0.995088	0.99505
32	0.995024	0.994992	0.994960	0.994928	0.994896	0.994864	0.994831	0.994799	0.994766	0.99473
33	0.994701	0.994668	0.994635	0.994602	0.994569	0.994536	0.994503	0.994470	0.994436	0.99440
34	0.994369	0.994336	0.994302	0.994268	0.994234	0.994201	0.994167	0.994132	0.994098	0.99406
35	0.994030	0.993995	0.993961	0.993926	0.993891	0.993857	0.993822	0.993787	0.993752	0.99371
36	0.993682	0.993647	0.993611	0.993576	0.993541	0.993505	0.993469	0.993434	0.993398	0.99336
37	0.993326	0.993290	0.993254	0.993218	0.993182	0.993146	0.993109	0.993073	0.993036	0.99300
38	0.992963	0.992926	0.992889	0.992852	0.992815	0.992778	0.992741	0.992704	0.992667	0.99262
39	0.992592	0.992554	0.992517	0.992479	0.992442	0.992404	0.992366	0.992328	0.992290	0.99225

Chapter 4

WATER VAPOR MIXING RATIO
AS A FLOW PARAMETER

INTRODUCTION

In various fluid flow systems the fluid of interest is moist air, i.e., a real-gas mixture of dry air and water vapor. There are several parameters that can be used to express the concentration of water vapor in the mixture. Of these, the water vapor mixing ratio, r, the ratio of the mass of water in the mixture to the mass of dry air in the mixture, will be discussed and a simple equation for the calculation of the saturation mixing ratio, r_s, will be developed. Several examples of the relationship between r and other flow parameters will be presented.

DENSITY OF MOIST AIR

The density of moist air is expressed[1] by

$$\rho = (PM_a/RTZ)[1 + (\epsilon - 1)e'/P] \tag{4.1}$$

where ρ is the density of moist air; P is the pressure, M_a is the apparent molecular weight of dry air;[1] R is the molar gas constant; T is absolute temperature; Z is the compressibility factor,[1] which accounts for the "nonideality" of the moist air; ϵ is the ratio of the molecular weight of water vapor to M_a; and e' is the effective saturation vapor pressure of water in moist air.

The quantity e' is expressed as

$$e' = (RH/100)fe_s = P/[1 + (\epsilon/r)] \tag{4.2}$$

where RH is relative humidity in percent; f, the enhancement factor, is a factor (near 1) which accounts for the fact that the saturation vapor pressure is greater than the saturation vapor pressure, e_s, of pure phase water vapor over a plane surface of pure ordinary liquid water;[1] and r is water vapor mixing ratio defined above.

In the absence of sources and sinks of water vapor, r is conserved. Therefore, under these conditions, r can be inferred from a measurement of relative humidity in a convenient region in a system (e.g., in a settling chamber in a flow meter calibration system or in the free stream of a wind tunnel) and used to characterize air in other regions of the system.

Substitution of Equation 4.2 in Equation 4.1 results in

$$\rho = (PM_a/RTZ)(1 + r)/[1 + (r/\epsilon)] \tag{4.3}$$

For convenience, we now use the meteorological quantity, virtual temperature (T'), given[2] by the expression

$$T' = T [1 + (r/\epsilon)]/(1 + r) \qquad (4.4)$$

and defined as "the temperature which dry air must have at the given barometric pressure p in order to have the same density as moist air at the same pressure p, and given temperature T, and mixing ratio, r, provided the dry and moist air behave in accordance with the perfect gas equation of state."[2]

Substitution of Equation 4.4 in Equation 4.3 results in

$$\rho = PM_a/RT'Z \qquad (4.5)$$

Equation 4.5 is a convenient form for calculating the density of moist air.

FREESTREAM AIR VELOCITY

Equation 4.5 is in a form that is useful in calculating the air velocity, V, along a streamline. A familiar form of the Bernoulli equation for calculating V is

$$V = [2\gamma P_0/(\gamma - 1)\rho_0]^{1/2} [1 - (1 - \Delta P/P_0)^{(\gamma - 1)/\gamma}]^{1/2} \qquad (4.6)$$

where γ is the real-gas ratio of specific heats for moist air; the subscript 0 indicates stagnation quantities; and $\Delta P = P_0 - P$, where P is static pressure. From Equation 4.5, $(P_0/\rho_0) = RT_0'Z/M_a$; thus, Equation 4.6 becomes

$$V = [2\gamma RT_0'Z/(\gamma - 1)M_a]^{1/2} [1 - (1 - \Delta P/P_0)^{(\gamma - 1)/\gamma}]^{1/2} \qquad (4.7)$$

REAL-GAS RATIO OF SPECIFIC HEATS, γ, FOR MOIST AIR

The real-gas ratio of specific heats, γ, for moist air also can be expressed in terms of r/ϵ, as

$$\gamma = [C_{pa} + (r/\epsilon)C_{pw}]/[C_{va} + (r/\epsilon)C_{vw}] \qquad (4.8)$$

where C_{pa} and C_{va} are the specific heats of dry air at constant pressure and volume, respectively; C_{pw} and C_{vw} are the corresponding quantities for water vapor.

CALCULATION OF SATURATION MIXING RATIO

Values of r are given by[3]

$$r = (RH/100)r_s \qquad (4.9)$$

where r_s is the saturation mixing ratio. Values of r_s are tabulated at various values of pressure and temperature in Reference 4, and in a more recent table[5] at discrete values of pressure and at temperature intervals of 10°C. Since the values in these two tables are in sufficiently good agreement, the values in Reference 4 will be used here to take advantage of the smaller temperature interval.

Values of r_s, in kg/kg, in the temperature range 17 to 37°C (at intervals of 1°C), and at intervals of 0.005 MPa in the pressure range 0.09 to 0.105 MPa (0.9 to 1.04 atmospheres) have been fitted by least squares to an equation quadratic in temperature t in °C:

$$r_s = a(P) + b(P) t + c(P) t^2 \qquad (4.10)$$

where P is pressure in MPa. As indicated, the P dependence has been assigned to the coefficients. The resulting equations, one for each value of P, are

$$r_s (0.0900 \text{ MPa}) = 1.47544 \times 10^{-2} - 8.13887 \times 10^{-4} t$$
$$+ 4.52355 \times 10^{-5} t^2 \qquad (4.11)$$

$$r_s (0.0950 \text{ Mpa}) = 1.38392 \times 10^{-2} - 7.56800 \times 10^{-4} t$$
$$+ 4.24506 \times 10^{-5} t^2 \qquad (4.12)$$

$$r_s (0.1000 \text{ MPa}) = 1.30112 \times 10^{-2} - 7.06064 \times 10^{-4} t$$
$$+ 3.99779 \times 10^{-5} t^2 \qquad (4.13)$$

$$r_s (0.1050 \text{ MPa}) = 1.22659 \times 10^{-2} - 6.60332 \times 10^{-4} t$$
$$+ 3.77519 \times 10^{-5} t^2 \qquad (4.14)$$

The values of the coefficients in Equations 4.11 through 4.14 were then fitted to equations linear in P. The resulting equations are

$$a(P) = 2.96378 \times 10^{-2} - 0.165846 P \qquad (4.15)$$

$$b(P) = -1.73137 \times 10^{-3} + 1.02266 \times 10^{-2} P \qquad (4.16)$$

$$c(P) = 8.99483 \times 10^{-5} - 4.98402 \times 10^{-4} P \qquad (4.17)$$

By inserting Equations 4.15 through 4.17 in Equation 4.10, the final equation for calculating r_s becomes

$$r_s = 2.96378 \times 10^{-2} - 0.165846\ P - 1.73137 \times 10^{-3}\ t$$

$$+ 1.02266 \times 10^{-2}\ Pt + 8.99483 \times 10^{-5}\ t^2$$

$$- 4.98402 \times 10^{-4}\ Pt^2 \qquad\qquad (4.18)$$

for P in MPa and t in °C.

UNCERTAINTY IN THE CALCULATION OF r_s

Values of r_s calculated using Equation 4.18 have been compared with the values used to develop the equation. The estimate of residual standard deviation (the estimate of the standard deviation of calculated minus data values) is 0.00015 kg/kg, which corresponds at the midpoint of the range of r_s to 0.51%. The contribution of the estimate of the residual standard deviation of r_s to the uncertainty in the calculation of ρ or P_0/ρ_0 is of the order of 0.001% and is therefore considered negligible for engineering calculations. The contribution to the uncertainty in the calculation of γ for moist air also is of the order of 0.001% and is therefore considered negligible for engineering calculations.

It is concluded that the relationships developed here can be useful in flow measurement or calibration calculations.

EXERCISE

1. For P = 0.101325 MPa and t = 23°C, what is the calculated value of r_s using Equation 4.18?

REFERENCES

1. **Jones, F. E.,** The air density equation and the transfer of the mass unit, *J. Res. Natl. Bur. Stand.*, 83, 419, 1978.
2. **Sheppard, P. A.,** The physical properties of air with reference to meteorological practices and the air-conditioning engineer, *Trans. ASTM*, 71, 915, 1949.
3. **List, R. J.,** *Smithsonian Meteorological Tables*, 6th rev. ed., Smithsonian Institution, Washington, DC, 1951, 348.
4. **List, R. J.,** *Smithsonian Meteorological Tables*, 6th rev. ed., Smithsonian Institution, Washington, DC, 1951, 304.
5. **Hyland, R. W. and Wexler, A.,** Formulations for the thermodynamic properties of dry air from 173.15 to 473.15, and of saturated moist air from 173.15 to 372.15 K, at pressures to 5 MPa, *Thermodynamic Properties of Dry Air, Moist Air and Water and SI Psychrometric Charts*, American Society of Heating, Refrigerating and Air-Conditioning Engineers, Inc., chap. 3.

Chapter 5

VISCOSITY OF GASES

INTRODUCTION

In the calibration and use of flow metering devices, the precise, accurate value of the viscosity of the gas being metered is often required. The availability of simple equations relating viscosity, μ, to temperature, t, and pressure, p, would be of convenience to the engineer or other worker in determining μ. In this chapter, simple interpolation formulas have been developed to enable one to conveniently make precise, accurate calculations of μ for dry air, nitrogen, carbon dioxide, helium, argon, and oxygen, using readily available hand-held calculators.[1]

The formulas are fitted to experimental data. The data were selected from the literature on the basis of claimed accuracy and precision, and of internal consistency. The last of these criteria is particularly important for the development of empirical equations for volume flow with μ as a dominant factor in the Poiseuille equation. The sets of data published by Kestin and his collaborators[2-11] meet these criteria and have been used to develop the formulas.

EXPERIMENTAL DATA

The ranges of temperature and pressure of interest in this chapter are $0°C \leq t \leq 50°C$ and $0 < p \leq 4$ MPa ($0 < p < 40$ atmospheres); 1 atmosphere (atm) = 101325 Pa = 0.101325 MPa = 14.69595 PSI = 760 mmHg. The experimental data used in developing the equations for μ are listed in Tables 5.1 through 5.6.

In Table 5.1 for dry air, the data in the first and third groups are taken from Table 3 of Reference 2; the second and fourth groups are taken from Table 2 of Reference 3; and the 50.00°C point is inferred from the data in Reference 3.

In Table 5.2 for nitrogen, the first and third groups are taken from Table 11 of Reference 2; the second, fifth, and sixth groups are taken from Table 8 of Reference 3; the fourth group is taken from Table 3 of Reference 4; the 20.58°C point is taken from Table 11 of Reference 2; and the 24.920°C point is taken from Table 3 of Reference 4.

In Table 5.3 for carbon dioxide, the 20.00°C data are taken from Table 5 of Reference 2; the second group is taken from Table 4 of Reference 4; and the 25.00°C point is taken from Table 3 of Reference 5.

In Table 5.4 for helium, the first group is taken from Table 7 of Reference 2; the 25.00°C point is taken from Table 4 of Reference 6; and the third group is taken from Table 5 of Reference 3.

In Table 5.5 for argon, the first and second groups are taken from Table 4 of Reference 2; the third group is taken from Table 4 of Reference 4; and the fourth group is taken from Table 3 of Reference 3.

TABLE 5.1
Viscosity of Dry Air

Temperature, t (°C)	Pressure, p (atm)	$\mu_{meas.}$ (µg/cm·s)	$\mu_{calc.}$ (µg/cm·s)	$\mu_{calc.} - \mu_{meas.}$ (µg/cm·s)
20.00	1.0	181.94	181.95	+0.01
20.00	4.39	182.38	182.37	−0.01
20.00	7.79	182.84	182.83	−0.01
20.00	11.22	183.32	183.32	0.00
20.00	14.61	183.88	183.84	−0.04
20.00	18.03	184.40	184.39	−0.01
20.00	21.40	184.98	184.96	−0.02
20.00	34.97	187.70	187.56	−0.14
23.44	1.07	183.75	183.85	+0.10
24.68	0.69	184.45	184.47	+0.02
24.00	1.71	184.33	184.22	−0.11
23.90	1.04	184.21	184.09	−0.12
23.89	0.83	184.13	184.06	−0.07
23.97	0.42	184.08	184.05	−0.03
25.00	1.00	184.62	184.68	+0.06
24.99	18.01	186.98	187.12	+0.14
24.99	35.12	190.18	190.32	+0.14
49.91	1.70	197.25	197.36	+0.11
49.39	1.08	196.99	197.04	+0.05
50.00[a]	1.00	197.31		

[a]   Used for calculating μ_1 only.

TABLE 5.2
Viscosity of Nitrogen

Temperature, t (°C)	Pressure, p (atm)	$\mu_{meas.}$ (µg/cm·s)	$\mu_{calc.}$ (µg/cm·s)	$\mu_{calc.} - \mu_{meas.}$ (µg/cm·s)
20.00	34.82	181.24	181.17	−0.07
20.00	20.48	178.64	178.56	−0.08
20.00	18.04	178.14	178.16	+0.02
20.00	14.59	177.58	177.61	+0.03
20.00	11.17	176.96	177.09	+0.13
20.00	7.79	176.50	176.60	+0.10
20.00	4.38	176.06	176.13	+0.07
20.00	1.00	175.69	175.69	0.00
20.00	0.687	175.62	175.65	+0.03
20.00	0.385	175.57	175.61	+0.04
21.75	1.57	176.52	176.54	+0.02
21.98	1.06	176.56	176.57	+0.01
24.44	1.67	177.80	177.77	−0.03
24.56	1.53	177.84	177.79	−0.05
24.72	1.43	177.86	177.85	−0.01

TABLE 5.2 (CONTINUED)
Viscosity of Nitrogen

Temperature, t (°C)	Pressure, p (atm)	$\mu_{meas.}$ (μg/cm·s)	$\mu_{calc.}$ (μg/cm·s)	$\mu_{calc.} - \mu_{meas.}$ (μg/cm·s)
24.84	1.32	177.86	177.90	+0.04
25.07	1.24	178.00	177.99	−0.01
25.15	1.15	177.99	178.02	+0.03
23.22	1.10	177.14	177.13	−0.01
23.15	1.04	177.05	177.08	+0.03
25.00	41.86	184.79	184.84	+0.05
25.00	28.21	182.12	182.16	+0.04
25.00	14.60	179.89	179.85	−0.04
25.00	7.80	178.83	178.84	+0.01
25.00	0.991	177.96	177.92	−0.04
25.00	1.000	177.98	177.93	−0.05
25.00	1.714	178.06	178.02	−0.04
25.00	2.300	178.13	178.09	−0.04
25.00	3.009	178.19	178.19	0.00
25.00	3.715	178.32	178.28	−0.04
25.00	4.341	178.41	178.37	−0.04
25.00	4.979	178.44	178.45	+0.01
25.00	5.579	178.59	178.53	−0.06
25.00	6.613	178.73	178.68	−0.05
25.00	7.307	178.81	178.77	−0.04
25.00	8.029	178.90	178.88	−0.02
25.00	8.954	179.04	179.01	−0.03
25.00	9.931	179.17	179.15	−0.02
25.00	12.567	179.59	179.54	−0.05
25.00	14.880	179.88	179.90	+0.02
25.00	20.051	180.66	180.73	+0.07
25.00	24.644	181.45	181.52	+0.07
25.00	29.838	182.49	182.46	−0.03
25.00	34.816	183.45	183.41	−0.04
25.00	39.919	184.43	184.44	+0.01
29.75	1.06	180.08	180.11	+0.03
29.93	1.06	180.12	180.20	+0.08
49.80	1.64	188.98	189.02	+0.04
49.73	1.64	190.01	189.98	−0.03
49.37	1.06	189.71	189.72	+0.01
20.58[a]	1.00	175.93		
24.920[a]	1.00	177.94		

[a] Used for calculating μ_1 only.

TABLE 5.3
Viscosity of Carbon Dioxide

Temperature, t (°C)	Pressure, p (atm)	$\mu_{meas.}$ (μg/cm·s)	$\mu_{calc.}$ (μg/cm·s)	$\mu_{calc.} - \mu_{meas.}$ (μg/cm·s)
20.00	18.02	148.62	148.61	−0.01
20.00	14.63	147.93	147.94	+0.01
20.00	11.18	147.39	147.41	+0.02
20.00	7.78	147.01	147.01	0.00
20.00	4.38	146.78	146.75	−0.03
20.00	0.976	146.63	146.63	0.00
20.00	0.640	146.56	146.62	+0.06
30.34	1.69	151.83	151.79	−0.04
30.40	1.06	151.81	151.81	0.00
49.12	1.08	161.75	161.75	0.00
25.00	1.00	149.09	149.09	0.00

TABLE 5.4
Viscosity of Helium

Temperature, t (°C)	Pressure, p (atm)	$\mu_{meas.}$ (μg/cm·s)	$\mu_{calc.}$ (μg/cm·s)	$\mu_{calc.} - \mu_{meas.}$ (μg/cm·s)
20.00	34.85	195.94	195.92	−0.02
20.00	21.40	195.99	196.02	+0.03
20.00	18.00	196.02	196.04	+0.02
20.00	14.61	196.09	196.07	−0.02
20.00	11.19	196.03	196.09	+0.06
20.00	7.80	196.17	196.11	−0.06
20.00	4.39	196.19	196.14	−0.05
20.00	1.01	196.14	196.16	+0.02
25.00	1.0	198.59	198.58	−0.01
24.01	1.57	198.12	198.10	−0.02
24.28	1.06	198.19	198.23	+0.04
30.49	1.71	201.20	201.17	−0.03
30.56	1.07	201.18	201.21	+0.03
53.00	1.61	211.16	211.15	−0.01
52.79	1.07	211.04	211.06	+0.02

In Table 5.6 for oxygen, the first and second groups are adjusted values from Table 12 of Reference 2; the third group is taken from Table 5 of Reference 11.

DEVELOPMENT OF EQUATIONS

The equations for $\mu(t,p)$, viscosity as function of temperature t in °C, and pressure p in various units were developed using a procedure[7] in which $\mu(t,p)$ is expressed as

TABLE 5.5
Viscosity of Argon

Temperature, t (°C)	Pressure, p (atm)	$\mu_{meas.}$ (µg/cm·s)	$\mu_{calc.}$ (µg/cm·s)	$\mu_{calc.} - \mu_{meas.}$ (µg/cm·s)
20.00	30.81	229.66	229.59	−0.07
20.00	21.42	227.16	227.17	+0.01
20.00	18.06	226.41	226.37	−0.04
20.00	14.63	225.58	225.59	+0.01
20.00	11.22	224.86	224.85	−0.01
20.00	7.79	224.18	224.14	−0.04
20.00	4.37	223.50	223.47	−0.03
20.00	1.00	222.86	222.85	−0.01
20.00	0.709	222.81	222.80	−0.01
20.00	0.384	222.73	222.74	+0.01
25.00	27.91	232.21	232.23	+0.02
25.00	21.41	230.42	230.60	+0.18
25.00	14.61	228.96	229.02	+0.06
25.00	7.81	227.57	227.58	+0.01
25.00	1.00	226.38	226.29	−0.09
25.00	1.000	226.34	226.29	−0.05
25.00	2.041	226.45	226.48	+0.03
25.00	3.034	226.70	226.66	−0.04
25.00	4.048	226.82	226.85	+0.03
25.00	5.110	227.13	227.05	−0.08
25.00	6.035	227.25	227.23	−0.02
25.00	7.158	227.49	227.45	−0.04
25.00	8.586	227.75	227.74	−0.01
25.00	10.029	228.12	228.04	−0.08
25.00	12.403	228.59	228.54	−0.05
25.00	15.288	229.13	229.17	+0.04
25.00	20.266	230.33	230.33	0.00
25.00	22.637	230.94	230.90	−0.04
25.00	24.542	231.36	231.38	+0.02
25.00	29.985	232.83	232.80	−0.03
25.00	35.156	234.26	234.24	−0.02
25.00	40.191	235.71	235.72	+0.01
24.33	1.56	225.83	225.93	+0.10
24.39	1.07	225.79	225.88	+0.09
49.37	1.73	242.68	242.93	+0.25
49.43	1.46	243.05	242.92	−0.13
49.83	1.22	243.17	243.15	−0.02
50.31	1.08	243.43	243.44	+0.01

TABLE 5.6
Viscosity of Oxygen

Temperature, t (°C)	Pressure, p (atm)	$\mu_{meas.}$ (μg/cm·s)	$\mu_{calc.}$ (μg/cm·s)	$\mu_{calc.} - \mu_{meas.}$ (μg/cm·s)
20.00	21.40	206.57	206.44	–0.13
20.00	18.02	205.86	205.82	–0.04
20.00	14.58	205.19	205.23	+0.04
20.00	11.19	204.70	204.67	–0.03
20.00	7.77	204.12	204.13	+0.01
20.00	4.39	203.61	203.64	+0.03
20.00	0.981	203.17	203.17	0.00
20.00	0.634	203.13	203.12	–0.01
25.00	41.83	213.87	213.88	+0.01
25.00	28.22	210.79	210.83	+0.04
25.00	14.61	208.25	208.28	+0.03
25.00	7.80	207.06	207.19	+0.13
25.00	0.998	206.25	206.22	–0.03
25.00	1	206.25	206.22	–0.03
35.00	1	212.18	212.18	0.00
50.00	1	220.80	220.79	–0.01

$$\mu(t,p) = \mu_0(t,0) + \Delta\mu(p) \tag{5.1}$$

where $\mu_0(t,0)$ is the viscosity at "zero pressure" and $\Delta\mu(p)$ is the experimental value of viscosity, $\mu_{meas.}$, minus $\mu_0(t,0)$. The procedure will be outlined below.

The units on μ in the development are μg/(cm s) (the use of the symbol μ for both viscosity and micro is unfortunate but not overpowering). The following conversion factors can be used to convert to other units:

 1 μg/(cm s) = 0.1 μPa·s

 1 μg/(cm s) = 1 μpoise

 1 μg/(cm s) = 0.0671969 × 10⁻⁶ lb/(ft s) = 0.0671969 μlb/(ft s)

AIR

The data pairs [20.00°C, 181.94 μg/(cm s)], [23.44°C, 183.75 μg/(cm s)], [23.90°C, 184.21 μg/(cm s)], [25.00°C, 184.62 μg/(cm s)], and [50.00°C, 197.31 μg/(cm s)], taken from Table 5.1, were fitted by least squares to an equation quadratic in t to enable calculation of μ at 0.101325 MPa (= 14.69595 PSI = 1 atmosphere), μ_1. The small departures of p from this value of presssure for these pairs are not significant. The resulting equation is

$$\mu_1 = 170.368 + 0.605434\ t - 1.33200 \times 10^{-3}\ t^2 \tag{5.2}$$

where μ_1 is in $\mu g/(cm\ s)$ and t is in °C.

To reduce μ_1 to "zero pressure", μ_0, the value of the increase in μ per 0.10325 MPa, 0.11 $\mu g/(cm\ s)$, estimated from the tabulated data, was subtracted from μ_1 resulting in

$$\mu_0 = 170.258 + 0.605434\ t - 1.33200 \times 10^{-3}\ t^2 \qquad (5.3)$$

The difference, $\Delta\mu$, between the experimental values of μ and calculated values of μ_0 was computed.

The values of $\Delta\mu$ were fitted by least squares to an equation quadratic in p. The resulting equation is

$$\Delta\mu = -2.44358 \times 10^{-3} + 1.17237\ p + 0.125541\ p^2 \qquad (5.4)$$

Equations 5.3 and 5.4 were added together to synthesize the final equation for calculating μ:

$$\mu_{calc.} = \mu_0 + \Delta\mu = 170.257 + 0.605434\ t - 1.33200 \times 10^{-3}\ t^2$$

$$+ 1.17237\ p + 0.125639\ p^2 \qquad (5.5)$$

where $\mu_{calc.}$ is in $\mu g/(cm\ s)$, t is in °C, and p is in MPa. Values of $\mu_{calc.}$ and differences between $\mu_{calc.}$ and experimental values, $\mu_{meas.}$, are listed in Table 5.1.

The estimate of residual standard deviation (RSD), i.e., the estimate of the standard deviation of ($\mu_{calc.} - \mu_{meas.}$), is 0.10 $\mu g/(cm\ s)$ for the 19 differences (n = 19). The estimate of the relative residual standard deviation (RRSD), i.e., the ratio of RSD to the mean $\mu_{meas.}$, is 0.05%. Kestin and Leidenfrost[2] estimated that for their measurements "a precision ranging from ±0.01% to ±0.07%, depending on the gas, has been achieved. The final accuracy of the measurements is estimated to be of the order of ±0.05%." DiPippo and Kestin[3] estimated the precision of their data to be ±0.05%; they reached the conclusion that "no meaningful assessment of the *accuracy* of the present data can be given if by accuracy we mean the *irreducible* discrepancy between the best measurements available at any particular time."

For μ in μpoise, t in °C, and p in PSI, Equation 5.5 becomes

$$\mu_{calc.} = 170.257 + 0.605434\ t - 1.33200 \times 10^{-3}\ t^2$$

$$+ 8.08321 \times 10^{-3}\ p + 5.97259 \times 10^{-6}\ p^2 \qquad (5.6)$$

For μ in μPa·s, t in °C, and p in MPa, Equation 5.5 becomes

$$\mu_{calc.} = 17.0257 + 6.05434 \times 10^{-2}\ t - 1.33200 \times 10^{-4}\ t^2$$

$$+ 0.117237\ p + 1.25639 \times 10^{-2}\ p^2 \qquad (5.7)$$

For μ in μlb/(ft s), t in °C, and p in PSI, Equation 5.5 becomes

$$\mu_{calc.} = 11.4407 + 4.06833 \times 10^{-2}\, t - 8.95063 \times 10^{-5}\, t^2$$

$$+ 5.43167 \times 10^{-4}\, p + 4.01340 \times 10^{-7}\, p^2 \qquad (5.8)$$

NITROGEN

The data pairs [20.58°C, 175.93 μg/(cm s)], [21.98°C, 176.56 μg/(cm s)], [23.15°C, 117.05 μg/(cm s)], [24.920°C, 177.94 μg/(cm s)], (29.75°C, 180.08 μg/cm s), and [49.37°C, 189.71 μg/(cm s)], taken from Table 5.2, were fitted by least squares to an equation quadratic in t to enable calculation of μ_1. The resulting equation is

$$\mu_1 = 167.335 + 0.392728\, t + 1.22474 \times 10^{-3}\, t^2 \qquad (5.9)$$

To reduce μ_1 to μ_0, 0.11 μg/(cm s) estimated from the tabulated data was subtracted from μ_1 resulting in

$$\mu_0 = 167.225 + 0.392728\, t + 1.22474 \times 10^{-3}\, t^2 \qquad (5.10)$$

The values of $\Delta\mu$ were fitted by least squares to an equation quadratic in p. The resulting equation is

$$\Delta\mu = -1.12860 \times 10^{-2} + 1.24165\, p + 9.87206 \times 10^{-2}\, p^2 \qquad (5.11)$$

Equations 5.10 and 5.11 were added together to synthesize the final equation for calculating μ:

$$\mu_{calc.} = 167.214 + 0.392728\, t + 1.22474 \times 10^{-3}\, t^2$$

$$+ 1.24165\, p + 9.87206 \times 10^{-2}\, p^2 \qquad (5.12)$$

where $\mu_{calc.}$ is in μg/(cm s), t is in °C, and p is in MPa.

The RSD is 0.05 μpoise for n = 50; the RRSD is 0.03%. Kestin et al.[4] estimated the accuracy of their experimental measurements to be of the order of ±0.2% and the relative precision to be 0.03%. The accuracy and precision estimates quoted above for air[2,3] apply also to nitrogen.

For μ in μpoise, t in °C, and p in PSI, Equation 5.12 becomes

$$\mu_{calc.} = 167.214 + 0.392728\, t + 1.22474 \times 10^{-3}\, t^2$$

$$+ 8.56087 \times 10^{-3}\, p + 4.69295 \times 10^{-6}\, p^2 \qquad (5.13)$$

For μ in μPa·s, t in °C, and p in MPa, Equation 5.12 becomes

$$\mu_{calc.} = 16.7214 + 3.92728 \times 10^{-2} \, t + 1.22474 \times 10^{-4} \, t^2$$

$$+ 0.124165 \, p + 9.87206 \times 10^{-3} \, p^2 \tag{5.14}$$

For μ in $\mu lb/(ft \, s)$, t in °C, and p in PSI, Equation 5.12 becomes

$$\mu_{calc.} = 11.2363 + 2.63901 \times 10^{-2} \, t + 8.22987 \times 10^{-5} \, t^2$$

$$+ 5.75264 \times 10^{-4} \, p + 3.15352 \times 10^{-7} \, p^2 \tag{5.15}$$

CARBON DIOXIDE

The data pairs [20.00°C, 146.63 $\mu g/(cm \, s)$], [25.00°C, 149.09 $\mu g/(cm \, s)$], [30.40°C, 151.81 $\mu g/(cm \, s)$], and [49.12°C, 161.75 $\mu g/(cm \, s)$], taken from Table 5.3, were fitted by least squares to an equation quadratic in t to enable calculation of μ_1. The resulting equation is

$$\mu_1 = 137.335 + 0.441133 \, t + 1.12987 \times 10^{-3} \, t^2 \tag{5.16}$$

To reduce μ_1 to μ_0, 0.13 $\mu g/(cm \, s)$ estimated from the tabulated data was subtracted from μ_1, resulting in

$$\mu_0 = 137.205 + 0.441133 \, t + 1.12987 \times 10^{-3} \, t^2 \tag{5.17}$$

The values of $\Delta\mu$ were fitted by least squares to an equation quadratic in p. The resulting equation is

$$\Delta\mu = 0.133827 + 6.28105 \times 10^{-2} \, p + 0.562974 \, p^2 \tag{5.18}$$

Equations 5.17 and 5.18 were added together to synthesize the final equation for calculating μ:

$$\mu_{calc.} = 137.339 + 0.441133 \, t + 1.12987 \times 10^{-3} \, t^2$$

$$+ 6.28105 \times 10^{-2} \, p + 0.562974 \, p^2 \tag{5.19}$$

where $\mu_{calc.}$ is in $\mu g/(cm \, s)$, t is in °C, and p is in MPa.

The RSD is 0.04 $\mu g/(cm \, s)$ for n = 11; the RRSD is 0.02%. Kestin et al.[5] estimated the accuracy of the 25.00°C point, "expressed as the maximum estimated uncertainty in the quoted values," to be ±0.1%. The accuracy and precision estimates quoted for air[2,3] apply also to carbon dioxide.

For $\mu_{calc.}$ in $\mu poise$, t in °C, and p in PSI, Equation 5.19 becomes

$$\mu_{calc.} = 137.339 + 0.441133 \, t + 1.12987 \times 10^{-3} \, t^2$$

$$+ 4.33063 \times 10^{-4} \, p + 2.67625 \times 10^{-5} \, p^2 \tag{5.20}$$

For μ in μPa·s, t in °C, and p in MPa, Equation 5.20 becomes

$$\mu_{calc.} = 13.7339 + 4.41133 \times 10^{-2}\, t + 1.12987 \times 10^{-4}\, t^2$$

$$+ 6.28105 \times 10^{-3}\, p + 5.62974 \times 10^{-2}\, p^2 \qquad (5.21)$$

For μ in μlb/(ft s), t in °C, and p in PSI, Equation 5.20 becomes

$$\mu_{calc.} = 9.22876 + 2.96428 \times 10^{-2}\, t + 7.59238 \times 10^{-5}\, t^2$$

$$+ 2.91005 \times 10^{-5}\, p + 1.79836 \times 10^{-6}\, p^2 \qquad (5.22)$$

HELIUM

The data pairs [20.00°C, 196.14 μg/(cm s)], [24.28°C, 198.19 μg/(cm s)], [25.00°C, 198.59 μg/(cm s)], [30.56°C, 201.18 μg/(cm s)], and [52.79°C, 211.04 μg/(cm s)] were fitted by least squares to an equation quadratic in t to enable calculation of μ_1. The effect of pressure on μ for helium is so small that no correction is made to reduce μ_1 to μ_0. The resulting equation is

$$\mu_1 = \mu_0 = 185.946 + 0.530773\, t - 1.04982 \times 10^{-3}\, t^2 \qquad (5.23)$$

The values of $\Delta\mu$ were fitted by least squares to an equation linear in p. The resulting equation is

$$\Delta\mu = 2.91664 \times 10^{-2} - 6.99813 \times 10^{-2}\, p \qquad (5.24)$$

Equations 5.23 and 5.24 were added together to synthesize the final equation for calculating μ:

$$\mu_{calc.} = 185.975 + 0.530773\, t - 1.04982 \times 10^{-3}\, t^2$$

$$- 6.99813 \times 10^{-2}\, p \qquad (5.25)$$

where $\mu_{calc.}$ is in μg/(cm s), t is in °C, and p is in MPa.

The RSD is 0.04 μg/(cm s) for n = 15; the RRSD is 0.02%. The accuracy and precision estimates quoted above for air[2,3] apply also to helium. The 25.00°C value is estimated to be accurate within ±0.1%.[6]

For μ in μpoise, t in °C, and p in PSI, Equation 5.25 becomes

$$\mu_{calc.} = 185.975 + 0.530773\, t - 1.04983 \times 10^{-3}\, t^2$$

$$- 4.82504 \times 10^{-4}\, p \qquad (5.26)$$

For μ in μPa·s, t in °C, and p in MPa, Equation 5.25 becomes

$$\mu_{calc.} = 18.5975 + 5.30773 \times 10^{-2}\, t - 1.04982 \times 10^{-4}\, t^2$$

$$- 6.99813 \times 10^{-3}\, p \tag{5.27}$$

For μ in μlb/(ft s), t in °C, and p in PSI, Equation 5.25 becomes

$$\mu_{calc.} = 12.4969 + 3.56663 \times 10^{-2}\, t - 7.05446 \times 10^{-5}\, t^2$$

$$- 3.24228 \times 10^{-5}\, p \tag{5.28}$$

ARGON

The data pairs [20.00°C, 222.86 μg/(cm s)], [24.39°C, 225.79 μg/(cm s)], [25.00°C, 226.36 μg/(cm s)], and [50.31°C, 243.43 μg/(cm s)], taken from Table 5.5, were fitted by least squares to an equation quadratic in t to enable calculation of μ_1. The resulting equation is

$$\mu_1 = 208.940 + 0.702190\, t - 3.30712 \times 10^{-4}\, t^2 \tag{5.29}$$

To reduce μ_1 to μ_0, 0.21 μg/(cm s) estimated from the tabulated data was subtracted from μ_1, resulting in

$$\mu_0 = 208.730 + 0.702190\, t - 3.30712 \times 10^{-4}\, t^2 \tag{5.30}$$

The values of $\Delta\mu$ were fitted by least squares to an equation quadratic in p. The resulting equation is

$$\Delta\mu = 3.17420 \times 10^{-2} + 1.73987\, p + 0.152358\, p^2 \tag{5.31}$$

Equations 5.30 and 5.31 were added together to synthesize the final equation for calculating μ:

$$\mu_{calc.} = 208.762 + 0.702190\, t - 3.30712 \times 10^{-4}\, t^2$$

$$+ 1.73987\, p + 0.152358\, p^2 \tag{5.32}$$

where $\mu_{calc.}$ is in μg/(cm s), t is in °C, and p is in MPa.

The RSD is 0.07 μg/(cm s) for n = 38; the RRSD is 0.03%. The estimates of accuracy and precision of the experimental data for argon are those quoted from References 2 through 4.

For μ in μpoise, t in °C, and p in PSI, Equation 5.32 becomes

$$\mu_{calc.} = 208.762 + 0.702190\, t - 3.30712 \times 10^{-4}\, t^2$$

$$+ 1.1996 \times 10^{-2}\, p + 7.24294 \times 10^{-6}\, p^2 \tag{5.33}$$

For μ in μPa·s, t in °C, and p in MPa, Equation 5.32 becomes

$$\mu_{calc.} = 20.8762 + 7.02190 \times 10^{-2}\, t - 3.30712 \times 10^{-5}\, t^2$$

$$+ 0.173987\, p + 1.52358 \times 10^{-2}\, p^2 \qquad (5.34)$$

For μ in μlb/ft·s, t in °C, and p in PSI, Equation 5.32 becomes

$$\mu_{calc.} = 14.0282 + 4.71850 \times 10^{-2}\, t - 2.22228 \times 10^{-5}\, t^2$$

$$+ 8.06094 \times 10^{-4}\, p + 4.86690 \times 10^{-7}\, p^2 \qquad (5.35)$$

OXYGEN

The data pairs [20.00°C, 203.17 μg/(cm s)], [25.00°C, 206.25 μg/(cm s)], [35.00°C, 212.18 μg/(cm s)], and [50.00°C, 220.80 μg/(cm s)], taken from Table 5.6, were fitted by least squares to an equation quadratic in t to enable calculation of μ_1. The resulting equation is

$$\mu_1 = 190.539 + 0.650043\, t - 8.97542 \times 10^{-4}\, t^2 \qquad (5.36)$$

To reduce μ_1 to μ_0, 0.12 μg/(cm s) estimated from the tabulated data was subtracted from μ_1, resulting in

$$\mu_0 = 190.419 + 0.650043\, t - 8.97542 \times 10^{-4}\, t^2 \qquad (5.37)$$

The values of $\Delta\mu$ were fitted by least squares to an equation quadratic in p. The resulting equation is

$$\Delta\mu = -0.0239595 + 0.130897\, p + 1.323418 \times 10^{-3}\, p^2 \qquad (5.38)$$

Equations 5.37 and 5.38 were added together to synthesize the final equation for calculating μ:

$$\mu_{calc.} = 190.395 + 0.650043\, t - 8.97542 \times 10^{-4}\, t^2$$

$$+ 1.29185\, p + 0.128975\, p^2 \qquad (5.39)$$

where $\mu_{calc.}$ is in μg/(cm s), t is in °C, and p is in MPa.

The RSD is 0.06 μg/(cm s) for n = 16; the RRSD is 0.03%. The estimates of accuracy and precision of the experimental data for oxygen are those quoted above for References 2 to 4.

For μ in μpoise, t in °C, and p in PSI, Equation 5.39 becomes

$$\mu_{calc.} = 190.395 + 0.650043\, t - 8.97542 \times 10^{-4}\, t^2$$

$$+ 8.90699 \times 10^{-3}\, p + 6.13118 \times 10^{-6}\, p^2 \qquad (5.40)$$

For μ in μPa·s, t in °C, and p in MPa, Equation 5.39 becomes

$$\mu_{calc.} = 19.0395 + 6.50043 \times 10^{-2}\, t - 8.97542 \times 10^{-5}\, t^2$$

$$+\ 0.129875\, p + 1.28975 \times 10^{-2}\, p^2 \qquad (5.41)$$

For μ in μlb/(ft s), t in °C, and p in PSI, Equation 5.39 becomes

$$\mu_{calc.} = 12.7940 + 4.36809 \times 10^{-2}\, t - 6.03120 \times 10^{-5}\, t^2$$

$$+\ 5.98522 \times 10^{-4}\, p + 4.11996 \times 10^{-7}\, p^2 \qquad (5.42)$$

RANGES OF APPLICATION OF THE EQUATIONS

The equations developed in this chapter are *interpolation* formulas fitted to experimental data. The range of strict application is indicated by the range of t and p of the experimental data listed in Tables 5.1 through 5.6.

In the absence of measurements of μ for temperatures below 20°C of comparable quality to that of the data in the tables, there are two options in extending calculations below 20° (0 < t < 20°C):

1. Apply the extended law of corresponding states.[4,8-11]
2. Extrapolate using the equations developed here, with probable loss in accuracy.

Either option might be followed until suitable experimental data at temperatures below 20°C became available.

TABLES

In the tables, the experimental data are given in the first three columns. The values of μ, $\mu_{calc.}$, calculated using the equations developed are listed in the fourth column. In the fifth column, the differences between calculated and measured μ, ($\mu_{calc.} - \mu_{meas.}$) are listed.

Temperatures in the tables are in °C, and pressures are in atmospheres (atm). The following conversion factors can be used to convert from atm to other units of pressure:

$$1\ atm = 101{,}325\ Pa$$

$$= 0.101325\ MPa$$

$$= 14.69595\ PSI$$

$$= 760\ mmHg$$

Viscosities in the tables are in μg/(cm s), which is equal to μpoise. The conversion factors to be used to convert from μg/(cm s) to other units of viscosity are listed above.

CONCLUSIONS

Equations (interpolation formulas fitted to experimental data) for the calculation of μ for dry air, nitrogen, carbon dioxide, helium, argon, and oxygen have been developed. The estimates of residual standard deviation for the fits are in close agreement with the estimates of precision for the experimental data in the above-stated ranges.

EXERCISES

1. Using the tables, what is calculated viscosity for dry air at 20°C and 1 atmosphere?
2. What is the maximum difference, in %, between measured and calculated viscosity for dry air?
3. What is the calculated viscosity of nitrogen at 20°C and 1 atmosphere?
4. What is the percentage difference between the calculated values of the viscosity of dry air and nitrogen at 20°C and 1 atmosphere?
5. What is the calculated value of the viscosity of argon at 20°C?
6. What is the difference between the calculated values of the viscosity of dry air and argon at 20°C and 1 atmosphere?

REFERENCES

1. **Jones, F. E.,** Interpolation Formulas for Viscosity of Air, Nitrogen, Carbon Dioxide, Helium, Argon, and Oxygen, NBS Technical Note 1186, February, 1984.
2. **Kestin, J. and Leidenfrost, W.,** An absolute determination of the viscosity of eleven gases, *Physica*, 25, 1033, 1959.
3. **DiPippo, R. and Kestin, J.,** The viscosity of seven gases up to 500°C and its statistical interpretation, in *Proc. 4th Symp. Thermophysical Properties,* American Society of Mechanical Engineers, College Park, MD, 1958, 304.
4. **Kestin, J., Paykoc, E., and Sengers, J. V.,** On the density expansion for viscosity in gases, *Physica*, 54, 1, 1971.
5. **Kestin, J., Ro, S. T., and Wakeham, W. A.,** Viscosity of carbon dioxide in the temperature range 25–700°C, *J. Chem. Phys.*, 58, 4114, 1972.
6. **Kestin, J., Khalifa, H. E., Ro, S. T., and Wakeham, W. A.,** The viscosity and diffusion coefficients of eighteen binary gaseous systems, *Physica*, 88A, 242, 1977.
7. **Kestin, J. and Whitelaw, J. H.,** The viscosity of dry and humid air, *Int. J. Heat Mass Transfer*, 7, 1245, 1964.
8. **Kestin, J., Ro, S. T., and Wakeham, W. A.,** An extended law of corresponding states for the equilibrium and transport properties of noble gases, *Physica*, 58, 165, 1972.

9. **Kestin, J., Ro, S. T., and Wakeham, W. A.,** Viscosity of the noble gases in the temperature range 25–700°C, *J. Chem. Phys.*, 56, 4119, 1970.
10. **Kestin, J. and Mason, E. A.,** Transport properties in gases (comparison between theory and experiment), in *AIP Conf. Proc. No. 11*, Kestin, J., Ed., 1973, 137.
11. **Hellemans, J. M., Kestin, J., and Ro, S. T.,** The viscosity of oxygen and some of its mixtures with other gases, *Physica*, 65, 362, 1973.

Chapter 6

GAMMA, RATIO OF SPECIFIC HEATS FOR AIR

INTRODUCTION

In this chapter, an equation for the calculation of the ratio of specific heats for dry air is developed. Also, equations for the calculation of the specific heats of water vapor and expressions for the specific heats of moist air are developed.

REAL-GAS SPECIFIC HEATS OF DRY AIR

The method of successive differences[1] has been used to relate the real-gas specific heat ratio, γ, the ratio of the specific heat at constant pressure to the specific heat at constant volume, for dry air to pressure, P. The resulting polylnomial is expressed as

$$\gamma (P,T) = a(T) + b(T)P + c(T)p^2 \qquad (6.1)$$

where T is air temperature in kelvins, K, and P is the pressure of the air in atmospheres (atm). As indicated, the T dependence is assigned to the coefficients a, b, and c. The coefficients are assumed to be quadratic in T:

$$a(T) = a_1 + a_2T + a_3T^2$$

$$b(T) = b_1 + b_2T + b_3T^2$$

$$c(T) = c_1 + c_2T + c_3T^2$$

By substituting these expressions for the coefficients, Equation 6.1 becomes

$$\gamma (P,T) = a_1 + a_2T + b_1P + b_2TP + a_3T_2 + c_1P^2 + c_2TP^2$$

$$+ b_3T^2P + c_3T^2P^2 \qquad (6.2)$$

The tabulated values of Hilsenrath et al.[2] in the range of P of 0.01 to 40 atmospheres and the range of T of 270 to 330K were fitted by nonlinear least squares. The resulting values of the coefficients and their approximate standard deviations are

$$a_1 = 1.39104, 4.2 \times 10^{-5}$$

$$a_2 = 8.58845 \times 10^{-5}, 2.8 \times 10^{-5}$$

$$b_1 = 1.41165 \times 10^{-2}, 1.1 \times 10^{-3}$$

$$b_2 = -6.81800 \times 10^{-5}, 7.5 \times 10^{-6}$$

$$a_3 = -1.87546 \times 10^{-7}, 4.7 \times 10^{-7}$$

$$c_1 = 5.33824 \times 10^{-5}, 2.7 \times 10^{-5}$$

$$c_2 = -3.06085 \times 10^{-7}, 1.8 \times 10^{-7}$$

$$b_3 = 9.00301 \times 10^{-8}, 1.2 \times 10^{-8}$$

$$c_3 = 4.35651 \times 10^{-10}, 3.1 \times 10^{-10}$$

The values of Hilsenrath et al.[2] of γ in the chosen ranges of T and P are tabulated in Table 6.1. The values of γ calculated using Equation 6.2 were compared with the values of Hilsenrath et al.[2]

The calculated values and the Table 6.1 values of γ were compared and the residuals (calculated — Table 6.1) were computed. The estimate of residual standard deviation (RSD) for 54 degrees of freedom is 9.0×10^{-5} which corresponds to a relative residual standard deviation (RRSD) of 0.0064%. Hilsenrath et al.[2] estimated that the uncertainty in their tabulated values of γ "does not exceed two or three parts in the last place tabulated except at the extremes where it may exceed two parts in the next to last place." One part in the last place tabulated would correspond to 0.007%; therefore, the random uncertainty in the interpolation formula, Equation 6.2, is comparable to the estimate of uncertainty in the tabulated values. Therefore, it can be concluded that the values of γ calculated using Equation 6.2 are in excellent agreement with the literature values.

TABLE 6.1
Values of γ for Dry Air from NBS Circular 564[2]

Pressure, P (atm)	Temperature, T (K)[a]						
	270	280	290	300	310	320	330
0.01	1.4006	1.4004	1.4002	1.4000	1.3997	1.3993	1.3990
0.1	1.4008	1.4006	1.4004	1.4001	1.3998	1.3995	1.3991
0.4	1.4014	1.4012	1.4009	1.4006	1.4003	1.4000	1.3995
0.7	1.4022	1.4018	1.4015	1.4012	1.4008	1.4004	1.3999
1	1.4029	1.4024	1.4020	1.4017	1.4013	1.4008	1.4004
4	1.4097	1.4087	1.4078	1.4070	1.4062	1.4053	1.4045
7	1.4166	1.4150	1.4135	1.4123	1.4111	1.4100	1.4089
10	1.4236	1.4214	1.4194	1.4177	1.4161	1.4111	1.4131
40	1.4956	1.4865	1.4786	1.4717	1.4658	1.4603	1.4553

[a]   At the writing of NBS Circular 564, °K = °C + 273.16.

The coefficients to be used in Equation 6.2 for T in °C and P in atm, γ (°C, atm); for T in °C and P in PSI, γ (°C, PSI); and for T in °C and P in megapascals γ (°C, MPa) are listed in Table 6.2.

REAL-GAS SPECIFIC HEATS OF WATER VAPOR

Keenan et al.[3,4] tabulated values of specific heats for water vapor, C_{pw}, and C_{vw}, at first in units of BTU (lb-mol)$^{-1}$ R^{-1} and then in J (g-mol)$^{-1}$ K^{-1}; R is a Rankine degree. C is specific heat, the subscripts p and v indicate constant pressure and constant volume, respectively, and the subscript w indicates water vapor. The data in the former units over the temperature range 250K to 333.33K were used in the present work to fit the specific heats to equations quadratic in T.

The fitted equation for C_{pw} is

$$C_{pw} = 8.1977 - 2.2324 \times 10^{-3}\,T + 5.5118 \times 10^{-6}\,T^2 \qquad (6.3)$$

for specific heat units of BTU (lb-mol)$^{-1}$ R^{-1} and T in kelvins (K).
For specific heat units of J (g-mol)$^{-1}$ K^{-1} and T in K,

$$C_{pw} = 34.320 - 9.3462 \times 10^{-3}\,T + 2.3076 \times 10^{-5}\,T^2 \qquad (6.4)$$

The fitted equation for C_{vw} is

$$C_{vw} = 6.2080 - 2.2180 \times 10^{-3}\,T + 5.5119 \times 10^{-6}\,T^2 \qquad (6.5)$$

for specific heat units of BTU (lb-mol)$^{-1}$ R^{-1} and T in K.
For specific heat units of J (g-mol)$^{-1}$ K^{-1} and T in K,

$$C_{vw} = 25.990 - 9.2859 \times 10^{-3}\,T + 2.3076 \times 10^{-5}\,T^2 \qquad (6.6)$$

TABLE 6.2
Coefficients to be Used in Equation 6.2 for Various Units of P and T

Coefficient	γ (atm, °K)	γ (atm, °C)	γ (PSI, °C)	γ (MPa, °C)
a_1	1.39104	1.40054	1.40054	1.40054
a_2	8.58807×10^{-5}	-1.65755×10^{-5}	-1.65755×10^{-5}	-1.65755×10^{-5}
b_1	1.41158×10^{-2}	2.21022×10^{-3}	1.50397×10^{-4}	2.18132×10^{-2}
b_2	-6.81782×10^{-5}	-1.18995×10^{-5}	-1.29252×10^{-6}	-1.87464×10^{-4}
a_3	-1.87546×10^{-7}	-1.87546×10^{-7}	-1.87546×10^{-7}	-1.87546×10^{-7}
a_1	5.33793×10^{-5}	2.27890×10^{-6}	1.05519×10^{-8}	2.21969×10^{-4}
c_2	-3.06076×10^{-7}	-6.80800×10^{-8}	-3.15228×10^{-10}	-6.63111×10^{-6}
b_3	9.00301×10^{-8}	9.00301×10^{-8}	6.12618×10^{-9}	8.88528×10^{-7}
c_3	4.35651×10^{-10}	4.35651×10^{-10}	2.01717×10^{-12}	4.24332×10^{-8}

The specific heats calculated using Equations 6.3 to 6.6 are within 1.2 parts in 10,000 of the values of Keenan et al.,[3,4] an excellent agreement.

γ OF MOIST AIR, γ_m

We now investigate γ for moist air. The molar specific heat at *constant pressure* of a mixture of dry air and water vapor, C_{pm}, can expressed as

$$C_{pm} = x_a C_{pa} + x_w C_{pw} \tag{6.7}$$

where x_a and x_w are the mole fractions of dry air and water vapor in the moist air mixture; the mole fraction of a species in a mixture is the ratio of the number of moles of the species to the total number of moles of the mixture. The subscript m indicates mixture, the subscript a indicates dry air, and the subscript w indicates water vapor.

C_{pa}, C_{pw}, C_{va}, and C_{vw} are the molar specific heats of dry air (a) and water vapor (w) at constant pressure (p) and constant volume (v).

Thus,

$$x_a + x_w = 1 \tag{6.8}$$

and

$$x_a = 1 - x_w \tag{6.9}$$

Equation 6.7 becomes

$$C_{pm} = (1 - x_w)C_{pa} + x_w C_{pw} \tag{6.10}$$

and

$$x_w = (m_w/M_w)/[(m_a/M_a) + (m_w/M_w)] \tag{6.11}$$

where m_a and m_w are the masses of air and water vapor, respectively, in the mixture and M_a and M_w are the molecular weights of dry air and water vapor, respectively.

From Chapter 2, $(M_w/M_A) = \epsilon$, and $(m_w/m_a) = r$. Therefore,

$$x_w = 1/[1 + (\epsilon/r)] \tag{6.12}$$

Then,

$$C_{pm} = \{1 - 1/[1 + (\epsilon/r)]\}C_{pa} + \{1/[1 + (\epsilon/r)]\}C_{pw} \tag{6.13}$$

$$= \{[1 + (\epsilon/r) - 1]C_{pa} + C_{pw}\}/[1 + (\epsilon/r)] \tag{6.14}$$

$$C_{pm} = [(\epsilon/r)C_{pa} + C_{pw}]/[1 + (\epsilon/r)] \qquad (6.15)$$

and

$$C_{pm} = [C_{pa} + (r/\epsilon)C_{pw}]/[1 + (r/\epsilon)] \qquad (6.16)$$

Similarly, the molar specific heat at *constant volume* for the mixture of dry air and water vapor, C_{vm}, is

$$C_{vm} = [C_{va} + (r/\epsilon)C_{vw}]/[1 + (r/\epsilon)] \qquad (6.17)$$

Therefore, γ_m, the ratio of C_{pm} to C_{vm}, is

$$\gamma_m = [C_{pa} + (r/\epsilon)C_{pw}]/[C_{va} + (r/\epsilon)C_{vw}] \qquad (6.18)$$

We now express (r/ϵ) in terms of relative humidity, RH. From the definition[5] of the effective vapor pressure in moist air, e',

$$r = (\epsilon + r)e'/P \qquad (6.19)$$

where P is the total pressure. From which

$$(r/\epsilon) = (e'/P)/(1 - e'/P) \qquad (6.20)$$

Jones[6] has shown that

$$e' = (RH/100)fe_s \qquad (6.21)$$

where RH is relative humidity in %; f (the enhancement factor) accounts for the fact that the effective saturation pressure of water in moist air is greater than the saturation vapor pressure of pure-phase water vapor over a plane surface of ordinary water, because the introduction of a second gas (air in this case) over the surface of the water increases the saturation concentration of water vapor above the surface of the water; and e_s is saturation water vapor pressure.

The ratio, (r/ϵ), thus becomes

$$(r/\epsilon) = [(RH/100)fe_s/P]/[1 - (RH/100)fe_s/P] \qquad (6.22)$$

Equation 6.22 can be inserted in Equations 6.15 to 6.18 to enable quantitative expression of the specific heats and their ratio.

As an example of the calculation of (r/ϵ), for a temperature of 23°C, $f = 1.0041$, $e_s = 2809$ Pa, an RH of 50%, and a total pressure, P, of 101325 Pa (1 atmosphere),

$$(RH/100)fe_s/P = (50/100)(1.0041)(2809)/101325$$

$$= 0.01392$$

and

$$[1 - (RH/100)fe_s/P = 0.9861$$

Then, from Equation 6.22,

$$(r/\epsilon) = 0.01392/0.9861 = 0.01412$$

It is interesting to examine (r/ϵ):

$$r = (\text{mass of water vapor/mass of dry air}) = n_wM_w/n_aM_a \qquad (6.23)$$

where the n's are numbers of moles and the M's are molecular weights.

$$\epsilon = (\text{molecular weight of water vapor/apparent molecular}$$
$$\text{weight of dry air}) = M_w/M_a \qquad (6.24)$$

Therefore,

$$(r/\epsilon) = (n_wM_w/n_aM_a)/(M_w/M_a) = n_w/n_a \qquad (6.25)$$

The ratio of specific heats for a mixture of dry air and water vapor is, thus,

$$\gamma_m = [C_{pa} + (r/\epsilon)C_{pw}]/[C_{va} + (r/\epsilon)C_{vw}] \qquad (6.26)$$

$$= [C_{pa} + (n_w/n_a)C_{pw}]/[C_{va} + (n_w/n_a)C_{vw}]$$

$$\gamma_m = [n_aC_{pa} + n_wC_{pw}]/[n_aC_{va} + n_wC_{vw}] \qquad (6.27)$$

Therefore, γ_m is equal to the ratio of the sums of the products of the numbers of moles and the corresponding molar specific heats, which is, in retrospect, the obvious result.

UNCERTAINTIES IN THE CALCULATION OF γ_m

We now investigate the variation of γ_m with the variation of various variables and parameters, leading to estimation of the uncertainties in the calculation of γ_m.

Differentiating Equation 6.18 first with respect to C_{Pa},

$$\partial\gamma_m/\partial C_{Pa} = 1/[C_{Va} + (r/\epsilon)C_{Vw}] \qquad (6.28)$$

Thus,

$$\Delta\gamma_m = \Delta C_{Pa}/[C_{Va} + (r/\epsilon)C_{Vw}] \qquad (6.29)$$

and

$$(\Delta\gamma_m/\gamma_m)_{Cpa} = \Delta C_{pa}/\gamma_m[C_{Va} + (r/\epsilon)C_{vw}] = \Delta C_{pa}/[C_{pa} + (r/\epsilon)C_{pw}] \quad (6.30)$$

where Δ represents variation or uncertainty.

Similarly,

$$(\Delta\gamma_m/\gamma_m)_{Cva} = -\Delta C_{va}/[C_{va} + (r/\epsilon)C_{vw}] \qquad (6.31)$$

$$(\Delta\gamma_m/\gamma_m)_{Cpw} = (r/\epsilon)\Delta C_{pw}/[C_{pa} + (r/\epsilon)C_{pw}] \qquad (6.32)$$

$$(\Delta\gamma_m/\gamma_m)_{Cvw} = -(r/\epsilon)\Delta C_{vw}/[C_{va} + (r/\epsilon)C_{vw}] \qquad (6.33)$$

$$(\Delta\gamma_m/\gamma_m)_{r/\epsilon} = \Delta(r/\epsilon)\{C_{pw}/[C_{pa} + (r/\epsilon)C_{pw}] - C_{vw}/[C_{va} + (r/\epsilon)C_{vw}]\} \quad (6.34)$$

Equations 7.30 to 7.34 express the relative uncertainties in γ due to uncertainties in C_{pa}, C_{va}, C_{pw}, C_{vw}, and (ϵ/r).

EXERCISES

1. Using Table 6.1, what is the value of γ for dry air at 290K and 1 atmosphere?
2. What is the value of γ at 300K and 1 atmosphere?
3. What is the percentage difference between the values of γ for 290K and 300K?

REFERENCES

1. **Booth, A. D.,** *Numerical Methods,* Butterworths, London, 1955, 7.
2. **Hilsenrath, J., Beckett, C. W., Fano, L., Hoge, H. J., Masi, J. F., Nuttall, R. L., Touloukian, Y. S.,** and **Woolley, H. W.,** Table of Thermal Properties of Gases, National Bureau of Standards, Circular 564, November 1955, 57.
3. **Keenan, J. H., Chao, J.,** and **Kaye, J.,** *Gas Tables,* 2nd ed., John Wiley & Sons, New York, 1980, 29.
4. **Keenan, J. H., Chao, J.,** and **Kaye, J.,** *Gas Tables, International Version,* 2nd ed., John Wiley & Sons, New York, 1983, 99.
5. **List, R. J.,** *Smithsonian Meteorological Tables,* 6th rev. ed., Smithsonian Institution, Washington, DC, 1951, 347.
6. **Jones, F. E.,** The air density equation and the transfer of the mass unit, *J. Res. Natl. Bur. Stand.,* 67C, 419, 1978.

Chapter 7

MEASUREMENT UNCERTAINTY

INTRODUCTION

A measurement of a quantity, volume flow rate, for example, is incomplete without a quantitative statement of the uncertainty of the measurement. In the past, the uncertainty of a measurement (the result of the application of a measurement process) has been considered to consist of random and systematic components.

The random component has generally been considered to be a measure of precision (or imprecision) of the measurement process as applied to the specific measurement. "The precision, or more correctly, the *im*precision of a measurement process is ordinarily summarized by the standard deviation of the process, which expresses the characteristic disagreement of repeated measurements of a single quantity by the process concerned, and thus serves to indicate how much a particular measurement is likely to differ from other values that the same measurement process might have provided in this instance, or might yield on remeasurement of the same quantity on another occasion."[1]

Eisenhart[1] has defined systematic error: "The systematic error, or bias, of a measurement process refers to its tendency to measure something other than what was intended; and is determined by the magnitude of the difference $\mu - \tau$ between the process average or limiting mean μ associated with measurement of a particular quantity by the measurement process concerned and the true value τ of the magnitude of this quantity."[1]

In this chapter, we discuss and apply the National Institute of Standards and Technology (NIST) guidelines for evaluating and expressing the uncertainty of measurement results.

NIST GUIDELINES

In October 1992, NIST instituted a new policy on expressing measurement uncertainty. The new policy was based on the approach to expressing uncertainty in measurement recommended by the International Committee for Weights and Measures (CIPM) and on the elaboration of that approach given in the *Guide to the Expression of Uncertainty in Measurement*[2] (hereafter referred to as the *Guide*). The *Guide* was prepared by individuals nominated by the International Bureau of Weights and Measures (BIPM), the International Electrochemical Commission (IEC), the International Organization for Standardization (ISO), or the International Organization of Legal Metrology (OIML). The CIPM approach is founded on Recommendation INC-1 (1980) of the Working Group on the Statement of Uncertainties.

NIST prepared a Technical Note, "Guidelines for Evaluating and Expressing the Uncertainty of NIST Measurement Results",[3] to assist in putting the policy into practice.

Since this treatment of uncertainty is in use in the U.S. and internationally, it is discussed here.

CLASSIFICATION OF COMPONENTS OF UNCERTAINTY

In the CIPM and NIST approach, the several components of the uncertainty of the result of a measurement may be grouped into two categories according to the method used to estimate the numerical values of the components:

Category A: those components that are evaluated by statistical methods.
Category B: those components that are evaluated by other than statistical methods.

Superficially, category A and B components resemble random and systematic components. The Guidelines[3] point out that this simple correspondence does not always exist, and that an alternative nomenclature to the terms "random uncertainty" and "systematic uncertainty" might be

"Component of uncertainty arising from a random effect"
"Component of uncertainty arising from a systematic effect"

STANDARD UNCERTAINTY

Each component of uncertainty is represented by an estimated standard deviation, referred to as a *standard uncertainty*. The standard uncertainty for a category A component is here given the symbol u_A. The evaluation of uncertainty by the statistical analysis of a series of observations is termed a *type A evaluation*.

A category B component of uncertainty, which may be considered an approximation to the corresponding standard deviation, is here given the symbol u_B. The evaluation of uncertainty by means other than statistical analysis of series of observations is termed a *type B evaluation*.

TYPE A EVALUATION OF STANDARD UNCERTAINTY

Type A evaluation of standard uncertainty may be based on any valid statistical method for treating data.

TYPE B EVALUATION OF STANDARD UNCERTAINTY

The NIST Guidelines[3] state that a type B evaluation of standard uncertainty is usually "based on scientific judgment using all the relevant information available, which may include

—previous measurement data,

—experience with, or general knowledge of, the behavior and property of relevant materials and instruments,

—manufacturer's specifications,

—data provided in calibration and other reports, and

—uncertainties assigned to reference data taken from handbooks."[3]

The Guidelines give examples and models for type B evaluations. In several of these, the measured quantity in question is modeled by a normal distribution with lower and upper limits –a and +b. For such a model "almost all" of the measured values of the quantity lie within plus and minus 3 standard deviations of the mean. Then a/3 can be used as an approximation of the desired standard deviation, u_B.

COMBINED STANDARD UNCERTAINTY

The *combined standard uncertainty*, with the symbol, u_C, of a measurement result is taken to represent the estimated standard deviation of the result. The combined standard uncertainty is obtained by taking the square root of the sum of the squares of the individual standard uncertainties, that is,

$$u_C = \sqrt{[(u_A)^2 + (u_B)^2]}$$

EXPANDED UNCERTAINTY

To define the interval about the measurement result within which the value of the quantity being measured is confidently believe to lie, the *expanded uncertainty* (with the symbol U) is intended to meet this requirement. The expanded uncertainty is obtained by multiplying u_C by a *coverage factor*, with the symbol k. Thus,

$$U = ku_C$$

Typically, k is in the range 2 to 3.

To be consistent with international practice, the value used by NIST for calculating U is, by convention, k = 2.

RELATIVE UNCERTAINTIES

The *relative standard uncertainty* is the ratio of the standard uncertainty, u_A or u_B, to the absolute value of the quantity measured.

The *relative combined uncertainty* is the ratio of the combined standard uncertainty, u_C, to the absolute value of the quantity measured.

The *relative expanded uncertainty* is the ratio of the expanded uncertainty, U, to the absolute value of the quantity measured.

Relative uncertainties can be expressed as percent or as decimals.

EXAMPLES OF DETERMINATION OF UNCERTAINTY

(1) For the calibration of a flow meter used to measure volume flow rate, the estimate of the random relative uncertainty (in the old terminology expressed as the estimate of standard deviation) was 0.05%. The estimate of relative systematic uncertainty was ±0.10%.

Formerly, the estimate of combined relative uncertainty would be

$$(3 \times 0.05) + 0.10 = 0.25\%$$

Using the NIST Guidelines and the same figures,

$$u_A = 0.05\%$$

$$u_B = 0.10\%/3 = 0.033\%$$

$$u_C = \sqrt{[(0.05)^2 + (0.033)^2]} = 0.06\%$$

For a coverage factor of $k = 2$,

$$U = ku_C = 2 \times 0.06\% = 0.12\%$$

(2) In the Guidelines,[3] an example of end-gauge calibration is given. Nine standard uncertainty component values are given: three of type A (5.8, 3.9, and 5.8 nanometers) and six of type B (25, 6.7, 1.7, 10.2, 2.9, and 16.6 nanometers).

Combining the type A standard uncertainty components by squaring each, summing, and taking the square root, the result is a type A standard uncertainty, u_A, of 9.1 nanometers. Combining the type B standard uncertainty components as were the type A standard uncertainty components, the result is a type B standard uncertainty, u_B, of 32.6 nanometers.

Combining the type A and type B standard uncertainties by squaring each, summing, and taking the square root, the result is a combined standard uncertainty, u_C, of 34 nanometers.

One could have combined all of the standard uncertainty components, disregarding type, and the same value of u_C would result. However, it is desirable to separate the uncertainties by type for many reasons including enabling the investigation of the possibility of reducing combined standard uncertainty.

(3) For the calibration of 1-kilogram mass standards, the estimate of the random uncertainty, expressed as the estimate of the standard deviation, was 0.01571 milligram. The estimate of the systematic uncertainty was 0.03000 milligram.

Formerly, the estimate of combined uncertainty would be

$$(3 \times 0.01571) + 0.03000 = 0.077 \text{ milligram}$$

Using the NIST Guidelines and the same figures,

$$u_A = 0.01571 \text{ milligram}$$

$$u_B = 0.03000/3 = 0.01000 \text{ milligram}$$

$$u_C = \sqrt{[(0.01571)^2 + (0.01000)^2]} = 0.01862 \text{ milligram}$$

For a coverage factor of $k = 2$,

$$U = 0.037 \text{ milligram}$$

REFERENCES

1. **Eisenhart, C. J.,** Realistic evaluation of the precision and accuracy of instrument calibration systems, *J. Res. Natl. Bur. Stand.*, 67C, 161, 1963.
2. *Guide to the Expression of Uncertainty in Measurement*, International Organization for Standards, Geneva, 1992.
3. **Taylor, B. N. and Kuyatt, C. E.,** Guidelines for evaluating and expressing the uncertainty of NIST measurement results, NIST Technical Note 1297, 1994 Edition, 1994.

Chapter 8

BUOYANCY CORRECTIONS IN WEIGHING

INTRODUCTION

Weighing on a balance essentially involves the balancing of forces. The action of the acceleration due to gravity on the weighed object generates a vertical force on the balance pan. The action of the acceleration due to gravity on the built-in weight or weights in the balance or on an external standard weight generates an offsetting force. In addition to these gravitational forces, the various objects and weights are partially supported by buoyant forces.

For accurate weighing, corrections accounting for the buoyant forces must be applied. It is the objective of this chapter to discuss buoyancy corrections and application of buoyancy corrections to flow calibration.

BUOYANT FORCE AND BUOYANCY CORRECTION

The *downward* vertical gravitational force exerted on a balance by an object is

$$F_g = Mg \tag{8.1}$$

where F_g is the gravitational force, M is the mass of the object, and g is the local acceleration due to gravity.

The net *upward* vertical force exerted on the body by the air in which the weighing is made is

$$F_b = \rho_a g(M/\rho_m) = M(\rho_a/\rho_m)g \tag{8.2}$$

where F_b is the force, ρ_a is the density of the air, and ρ_m is the density of the object. This net upward vertical force can be called the *buoyant force*.

The overall *net* vertical force, F, is

$$F = F_g - F_b = M[1 - (\rho_a/\rho_m)]g \tag{8.3}$$

The quantity $[1 - \rho_a/\rho_m)]$ when applied to weighing is a *buoyancy correction factor*. The net force exerted on the balance is thus *less* than the gravitational force.

For a reference air density, ρ_a, of 0.0012 g/cm³ and an object density, ρ_m, of 8.0 g/cm³ (approximating the density of stainless steel),

$$(\rho_a/\rho_m) = 0.00015$$

and the buoyancy correction factor is

$$[1 - (\rho_a/\rho_m)] = 0.99985$$

The *buoyancy correction*, $(\rho_a/\rho_m)M$, 0.00015 M, corresponds to 150 parts per million (ppm) of M or 0.015% of M.

If water, with an approximate density of 1.0 g/cm^3, were being weighed, the buoyancy correction would be 1200 parts per million of M or 0.12% of M.

For the nominal stainless steel case, the correction for a mass of 1 kilogram would be 150 milligrams; for a mass of 100 grams, the correction would be 15 milligrams. For the water case, the correction for a mass of 1 kilogram would be 1.2 grams; for a mass of 100 grams, the correction would be 120 milligrams.

For a reference air density, ρ_0, of 0.0012 g/cm^3 and substance densities from 0.7 to 22 g/cm^3, the values of the buoyancy correction factor range from 0.9982857 to 0.9999455, and the values of the buoyancy correction range from 1714.3 to 54.5 ppm. Values of the buoyancy correction factor and the ratio ρ_a/ρ_m are tabulated in Table 8.1.

The significance of these corrections depends on the desired accuracy for the particular substance in the particular situation and on the precision of the balance, among other things.

APPLICATION OF THE SIMPLE BUOYANCY CORRECTION FACTOR TO WEIGHING ON A SINGLE-PAN ANALYTICAL BALANCE

In a simple case using a single-pan analytical balance, the unknown mass of an object, M_x, is balanced by built-in weights of total mass, M_s. The balance indication, U_s, is equal to M_s. It is assumed that the built-in weights exactly balance M_x.

The force exerted on the balance pan by the object X of mass M_x is

$$F_x = M_x(1 - \rho_a/\rho_x)g \qquad (8.4)$$

where ρ_a is the density of air and ρ_x is the density of the weight (object X).

The force exerted on the balance pan by an assemblage of built-in weights of total mass M_s is

$$F_s = M_s(1 - \rho_a/\rho_s)g \qquad (8.5)$$

where ρ_s is the density of the assemblage of built-in weights.

When these forces are equal,

$$F_x = F_s \qquad (8.6)$$

$$M_x(1 - \rho_a/\rho_x) = M_s(1 - \rho_a/\rho_s) \qquad (8.7)$$

$$M_x = M_s(1 - \rho_a/\rho_s)/(1 - \rho_a/\rho_x) \qquad (8.8)$$

TABLE 8.1
Buoyancy Correction Factors and Ratios
for $\rho_a = 0.0012$ g/cm^3

$$A = [1 - (\rho_a/\rho_m)], \, B = (\rho_a/\rho_m)$$

ρ_m (g/cm^3)	A	B		
		%	ppm	mg/100g
0.7	0.998286	0.1714	1714	171.4
1.0	0.9988	0.12	1200	120.0
1.5	0.9992	0.08	800	80.0
2.0	0.9994	0.06	600	60.0
3.0	0.9996	0.04	400	40.0
4.0	0.9997	0.03	300	30.0
5.0	0.99976	0.024	240	24.0
6.0	0.9998	0.020	200	20.0
7.0	0.999829	0.0171	171	17.1
8.0	0.99985	0.015	150	15.0
9.0	0.999867	0.0133	133	13.3
10.0	0.99988	0.012	120	12.0
11.0	0.999891	0.0109	109	10.9
12.0	0.9999	0.01	100	10.0
13.0	0.999908	0.0092	92	9.2
14.0	0.999914	0.0086	86	8.6
15.0	0.99992	0.0080	80	8.0
16.0	0.999925	0.0075	75	7.5
16.5	0.999927	0.0073	73	7.3
17.0	0.999929	0.0071	71	7.1
18.0	0.999933	0.0067	67	6.7
19.0	0.999937	0.0063	63	6.3
20.0	0.999940	0.0060	60	6.0
21.0	0.999943	0.0057	57	5.7
22.0	0.999945	0.0055	55	5.5

We now make calculations of M_x for several values of ρ_x and for the following fixed values:

$$\rho_a = 0.0012 \text{ g/cm}^3$$

$$t = t_0 = 20°C$$

$$\rho_s = 8.0 \text{ g/cm}^3$$

$$M_s = 1000 \text{ g}$$

For $\rho_x = 1.0$ g/cm^3, the approximate density of water,

$$M_x = 1000 \, (1 - 0.0012/8.0)/(1 - 0.0012/1.0)$$

$$M_x = 1001.051 \text{ g}$$

That is, under these conditions, *1001.051 g of weight X of density 1.0 g/cm³* would balance 1000 g of S weights of density 8.0 g/cm³. The difference, 1.051 g, between the X and S weights is due to the difference in buoyant forces acting on the weights. This, of course, is due to the difference in density of the weights and, consequently, to the difference in volume of air displaced by the weights.

For $\rho_x = 2.7 \text{ g/cm}^3$,

$$M_x = 1000 \ (1 - 0.0012/8.0)/(1 - 0.0012/2.7)$$

$$M_x = 1000.295 \text{ g}$$

For $\rho_x = 16.6 \text{ g/cm}^3$,

$$M_x = 1000 \ (1 - 0.0012/8.0)/(1 - 0.0012/16.6)$$

$$M_x = 999.922 \text{ g}$$

For $\rho_x = 22.0 \text{ g/cm}^3$,

$$M_x = 1000 \ (1 - 0.0012/8.0)/(1 - 0.0012/22.0)$$

$$M_x = 999.905 \text{ g}$$

We note that, for these latter two cases, M_x is less than 1000 g. This is, of course, because ρ_x is greater than 8.0 g/cm³ in these two cases.

For $\rho_x = 8.0 \text{ g/cm}^3$,

$$M_x = 1000 \ (1 - 0.0012/8.0)/(1 - 0.0012/8.0)$$

$$M_x = 1000 \text{ g}$$

For $\rho_x = 8.0$ g/cm³, we have the conditions that define apparent mass.

THE ELECTRONIC ANALYTICAL BALANCE

In an electronic force balance

1. An electronic force is generated to oppose the net gravitational and buoyant force imposed by the mass being weighed.
2. The readout of the balance is proportional to the current in a servomotor coil.

3. In calibration of the balance, a calibrating weight is used and the electronic circuitry is adjusted so that the readout indicates the *apparent mass* of the calibrating weight.

ELECTRONIC BALANCE CALIBRATION AND USE

We now investigate the calibration and performance of an electronic balance with a built-in calibrating weight.

The *true mass* of the calibrating weight is assumed to be *100 g*. Throughout this discussion, M with no superscript refers to true mass. The density of the calibrating weight, ρ_b, is assumed to be *8.0 g/cm³*, ρ_r. The temperature at which the balance is calibrated at the factory is assumed to be *20°C*, and the air density at the factory is assumed to be *0.0012 g/cm³*.

The defining equation for *apparent mass* is

$$^AM_x = {}^TM_r = {}^TM_x\,(1 - \rho_0/\rho_x)/(1 - \rho_0/\rho_r) \tag{8.9}$$

where ^AM_x is the apparent mass of the object of interest; ^TM_r is the true mass of a quantity of reference material of density 8.0 g/cm³; ^TM_x is the true mass of the object; ρ_0 is the reference air density, 0.0012 g/cm³; ρ_x is the density of the object; and ρ_r is the density of the reference material, 8.0 g/cm³.

Under the above conditions,

$$^TM_x = 100 \text{ g} = {}^TM_b$$

$$\rho_0 = 0.0012 \text{ g/cm}^3$$

$$\rho_r = 8.0 \text{ g/cm}^3 = \rho_b$$

$$^TM_r = 100 \text{ g}$$

$$\rho_x = 8.0 \text{ g/cm}^3$$

where ^TM_b and ρ_b are the true mass and density, respectively, of the calibrating or built-in weight.

Note that "mass" is a property and "weight" is an object.

Thus,

$$^AM_b = 100[1 - (0.0012/8.0)]/[1 - (0.0012/8.0)] = {}^TM_b = 100 \text{ g} \tag{8.10}$$

Therefore, at the factory under the above conditions, the apparent mass of the calibrating weight is equal to the true mass of the calibrating weight.

At the *factory*, the balance is calibrated (in this case, using the built-in calibrating weight) to indicate *apparent mass*.

In the balance, an electromotive force, F, is generated to equal and oppose the net force impressed on the balance pan by the gravitational force minus the buoyant force. The electromotive force, F, is generated by the current, I, passing through the coil of an electromotive force cell. F is proportional to I. The indication of the balance, U, is proportional to I at equilibrium. Thus,

$$F = kI \tag{8.11}$$

$$U = cI \tag{8.12}$$

where k and c are constants of proportionality.

$$I = F/k = U/c \tag{8.13}$$

$$U = (c/k)F = KF \tag{8.14}$$

where $(c/k) = K$.

Again, at the *factory*,

$$\rho_a = \rho_0 = 0.0012 \text{ g/cm}^3$$

$$g = g_f = \text{the acceleration due to gravity at the balance location in the factory}$$

$$\rho_b = \rho_r = 8.0 \text{ g/cm}^3$$

$$t = t_0 = 20°C$$

$$1/K = K_0$$

The force, F_x, exerted on the balance by an unknown mass, M_x, is

$$F_x = M_x(1 - \rho_0/\rho_x)g_f = U_x K_0 \tag{8.15}$$

The force, F_b, exerted on the balance by the built-in calibrating weight of mass M_b and density 8.0 g/cm³ is

$$F_b = M_b(1 - \rho_0/\rho_b)g_f = U K_0 \tag{8.16}$$

At the factory, the electronics are adjusted in such a way that the indication of the balance, U, is equal to the apparent mass of the built-in weight (e.g., 100 g) with the built-in weight introduced to the balance. We shall refer to this operation as the *adjusting* of the balance rather than the *calibration* of the balance.

Thus, under the above conditions,

$$U = {}^AM_b = {}^TM_b \tag{8.17}$$

Then F_b is given by

$$F_b = U(1 - \rho_0/\rho_b)g_f = UK_0 \tag{8.18}$$

and, thus,

$$K_0 = (1 - \rho_0/\rho_b)g_f \tag{8.19}$$

At the *factory*, for a *standard* weight of true mass M_s, and the conditions:

$$\rho_a = \rho_0 = 0.0012 \text{ g/cm}^3$$

$$\rho_s$$

$$\rho_b = 8.0 \text{ g/cm}^3$$

$$t_0 = 20°C$$

$$g = g_f$$

$$1/K = K_0$$

the force exerted on the balance is

$$F_s = M_s(1 - \rho_0/\rho_s)g_f = U_sK_0 = U_s(1 - \rho_0/\rho_b)g_f \tag{8.20}$$

where U_s is the balance indication with M_s on the balance. Then,

$$M_s = U_s(1 - \rho_0/\rho_b)/(1 - \rho_0/\rho_s) \tag{8.21}$$

and

$$U_s = M_s(1 - \rho_0/\rho_s)/(1 - \rho_0/\rho_b) \tag{8.22}$$

This last equation is recognized to be the definition of the apparent mass of the standard weight, M_s, since $\rho_b = \rho_r$. Therefore, the indication of the balance is the apparent mass of the standard weight:

$$U_s = {}^AM_s$$

Similarly, for a weight of unknown true mass M_x,

$$U_x = M_x(1 - \rho_0/\rho_x)/(1 - \rho_0/\rho_b) \qquad (8.23)$$

and, again, the balance indication is equal to the apparent mass of the weight on the pan:

$$U_x = {}^AM_x$$

USUAL CASE FOR WHICH THE AIR DENSITY IS NOT THE REFERENCE VALUE

We now consider the more usual case for which the air density, ρ_a, is not equal to the reference value, $\rho_0 = 0.0012$ g/cm^3.

In the *laboratory* for adjusting of the balance using the built-in weight and the conditions:

$$\rho_a = \rho_a$$

$$t_0 = 20°C$$

$$g = g_L$$

$$\rho_b = 8.0 \text{ g/cm}^3$$

where g_L is the local acceleration of gravity in the laboratory at the location of the balance, the force exerted on the balance by the introduction of the built-in weight of mass M_b is

$$F_b = M_b(1 - \rho_a/\rho_b)g_L = U_b K_L \qquad (8.24)$$

The electronics of the balance are adjusted so that the scale indication, U_b, is equal to the apparent mass of the built-in weight (which is also the true mass). Then

$$K_L = (1 - \rho_a/\rho_b)g_L \qquad (8.25)$$

For a *standard* weight of true mass M_s on the pan of the balance, the force exerted by the standard weight is

$$F_s = M_s(1 - \rho_a/\rho_s)g_L = U_s(1 - \rho_a/\rho_b)g_L \qquad (8.26)$$

and

$$U_s = M_s(1 - \rho_a/\rho_s)/(1 - \rho_a/\rho_b) \qquad (8.27)$$

and

$$M_s = U_s(1 - \rho_a/\rho_b)/(1 - \rho_a/\rho_s) \tag{8.28}$$

Therefore, if the balance were operating perfectly, with the standard weight on the pan, the balance indication would be equal to the right side of Equation 8.27. Deviation of the balance indication from this value would represent a weighing error or a random deviation.

For a weight of unknown mass, M_x, on the balance, the force exerted on the balance is

$$F_x = M_x(1 - \rho_a/\rho_x)g_L = U_x(1 - \rho_a/\rho_b)g_L \tag{8.29}$$

and

$$U_x = M_x(1 - \rho_a/\rho_x)/(1 - \rho_a/\rho_b) \tag{8.30}$$

and

$$M_x = U_x(1 - \rho_a/\rho_b)/(1 - \rho_a/\rho_x) \tag{8.31}$$

Therefore, if the balance were operating perfectly, with the unknown weight on the pan, the balance indication would be equal to the right side of Equation 8.30, and the true mass of the unknown would be calculated using Equation 8.31.

Note that it is the *true mass* of the unknown that is the desired mass quantity, *not* the *indication* of the balance, and *not* the *apparent mass* of the unknown. Even if the apparent mass were measured perfectly, a calculation must be made to determine the true mass.

EXAMPLES OF EFFECTS OF FAILURE
TO MAKE BUOYANCY CORRECTIONS

In the calibration of flow meters for liquids at the National Institute of Standards and Technology (NIST), two liquids are used: (1) water and (2) Stoddard's solvent (Mil. Spec. 7024, type 2).

The density of Stoddard's solvent at 60°F (15.56°C) is 770.9 kg/m³ or 0.7709 g/cm³. For an air density, ρ_a, of 0.0012 g/cm³, a mass of solvent, M_x, of 1000 g, a density of the balance built-in weight of 8.0 g cm³, and a density of Stoddard's solvent of 0.7709 g cm³, the indication of the balance using Equation 8.30 is

$$U_x = 1000 \, (1 - 0.0012/0.7709)/(1 - 0.0012/8.0)$$

U_x is then 998.593 g.

If the indication of the balance, 998.593 g, were taken to be the measurement of the mass of the solvent, it would be in error by 1.407 g (1000 – 998.593) or 0.1407%, a quite significant error. This is the consequence of not making a buoyancy correction. The true value of the mass of solvent would be calculated by applying the buoyancy correction, that is, by dividing the balance indication by the buoyancy correction factor, [(1 – 0.0012/0/7709)/(1 – 0.0012/8.0)]:

$$M_x = 998.593/[(1 - 0.0012/0.7709)/(1 - 0.0012/8.0)]$$

$$M_x = 1000 \text{ g}$$

The density of air-saturated water at 60°F (15.56°C) is 999.010 kg/m³ or 0.999010 g/cm³ (Chapter 3). For the other conditions in the Stoddard's solvent example, Equation 8.30 becomes

$$U_x = 1000 \ (1 - 0.0012/0.999010)/(1 - 0.0012/8.0)$$

U_x is then 998.949 g.

If the indication of the balance, 998.949 g, were taken to be the measurement of the mass of the water, it would be in error by 1.051 g (1000 – 998.949) or 0.1051%, again a quite significant error. Again, this is the consequence of not making a buoyancy correction. The true value of the mass of water would be calculated by dividing the balance indication by the buoyancy correction factor, [(1 – 0.0012/0.999010)/(1 – 0.0012/8.0)]:

$$M_x = 998.949/[(1 - 0.0012/0.999010)/(1 - 0.0012/8.0)]$$

$$M_x = 1000 \text{ g}$$

CONCLUSION

Failure to make buoyancy corrections can result in quite significant errors in flow-related calibrations and measurements.

EXERCISES

1. For

$$M = 100 \text{ g}$$

$$\rho_a = 0.0012 \text{ g/cm}^3$$

$$\rho_m = 1.0 \text{ g/cm}^3$$

(a) Calculate the simple buoyancy correction factor.
(b) Calculate the simple buoyancy correction, in g.

2. For

$$M_x = 100 \text{ g}$$

$$\rho_a = 0.0012 \text{ g/cm}^3$$

$$\rho_x = 1.0 \text{ g/cm}^3$$

(a) What is the volume of the object?
(b) What is the mass of the same volume of air, that is, the mass of the volume of air displaced by the object?
(c) What is the ratio of the mass of air to the mass of the object, M_x, in %?

3. For

$$M = 100 \text{ g}$$

$$\rho_a = 0.0012 \text{ g/cm}^3$$

$$\rho_m = 1.0 \text{ g/cm}^3$$

(a) Calculate how much of the mass, in g, is supported by the balance pan.
(b) Calculate how much of the mass, in g, is supported by the air.

Chapter 9

REAL-GAS CRITICAL FLOW FACTOR, C*

INTRODUCTION

The real-gas critical-flow factor, C*, is defined as

$$C^* = G \, (RT_0)^{1/2}/P_0 \tag{9.1}$$

where G is the mass flow rate per unit area, R is the specific gas constant (the ratio of the universal gas constant to the molecular weight of the gas), T_0 is the stagnation temperature, and P_0 is the stagnation pressure. C* is dimensionless.
Johnson[1] calculated and tabulated C* for critical-flow nozzles.

INTERPOLATION FORMULA FOR C*

In this chapter, an interpolation formula for C* for the pressure range 0 to 40 atmospheres (587.838 PSI) and the temperature range 269.44 to 333.33K (–3.71 to 60.18°C) is developed.
The relationship between C*, P, and T is expressed as

$$C^*(P,T) = a(T) + b(T)P + c(T)P^2 \tag{9.2}$$

where

$$a(T) = a_1 + a_2T + a_3T^2$$

$$b(T) = b_1 + b_2T + b_3T^2$$

$$c(T) = c_1 + c_2T + c_3T^2$$

Substitution of a(T), b(T), and c(T) in Equation 9.2 results in

$$C^* = a_1 + a_2T + a_3T^2 + b_1P + b_2PT + b_3PT^2$$

$$+ [c_1P^2 + c_2P^2T + c_3P^2T^2] \tag{9.3}$$

A preliminary analysis of the tabulated values indicated that the terms in the brackets in Equation 9.3 are probably not significant.
The tabulated[1] values of C* in the ranges of P and T of interest were fitted to an equation of the form of Equation 9.3, without and with the terms in the brackets. The fitted equation with first six terms is

$$C^* = A + Bt + CP + DtP + Et^2 + HPt^2 \tag{9.4}$$

where the coefficients are

$$A = 0.684845$$

$$B = 1.03462 \times 10^{-6}$$

$$C = 2.54266 \times 10^{-5}$$

$$D = -3.09313 \times 10^{-7}$$

$$E = -5.06917 \times 10^{-8}$$

$$H = 1.48528 \times 10^{-9}$$

for t in °C and P in PSI.

Values of C* calculated using Equation 9.4 were compared with the corresponding tabulated literature values.[1] The residual is the calculated value minus the tabulated value. The estimate of the standard deviation of the residuals was 0.000038. The estimate of standard deviation assigned to roundoff in the tabulated values is 0.000029. Therefore, the imprecision of calculated values is comparable to the estimate of standard deviation due to roundoff.

Johnson[1] tabulated values of C* to four digits, "carried to one physically nonsignificant figure."[1] The uncertainty in the tabulated values was thus much larger than the estimate of standard deviation of the residuals. Equation 9.4 is thus seen to be a very efficient, simple interpolation equation. The ratio of the estimate of the standard deviation of the residuals to C* is 0.0056%, an excellent result.

EXERCISES

1. What is the calculated value of C* for P = 1 atm and T = 296.15K?
2. What is the calculated value of C* for pressure = 14.69595 PSI and temperature = 23°C?

REFERENCE

1. **Johnson, R. C.,** Real-Gas Effects in Critical-Flow-Through Nozzles and Tabulated Thermodynamic Properties, NASA Technical Note NASA TN D-2565, 1965, 36.

Chapter 10

SUBSONIC FLOW AND DISCHARGE COEFFICIENTS

INTRODUCTION

Bernoulli's Principle for incompressible flow (in which changes in density are considered to be negligible) can be expressed by

$$V^2/2 + P/\rho + gZ = \text{constant} \qquad (10.1)$$

where V is velocity, P is pressure, ρ is density, g is the acceleration due to gravity, and Z is height above an arbitrary datum plane. For compressible flow, the second term in Equation 10.1 would be replaced by $\int dP/\rho$.

SUBSONIC FLOW OF GAS THROUGH VENTURIS, NOZZLES, AND ORIFICES

For level flow (with gravitational effects negligible),

$$V^2/2 + P/\rho = \text{constant} \qquad (10.2)$$

$$V_1^2/2 + P_1/\rho = V_2^2/2 + P_2/\rho \qquad (10.3)$$

where the subscripts 1 and 2 refer to upstream and downstream, respectively, of the venturi, nozzle, or orifice.

The equation of continuity is

$$\dot{\rho} + \text{div}\,(\rho V) = 0 \qquad (10.4)$$

where $\dot{\rho}$ is the time rate of change of ρ, the second term is the divergence of ρV, and the bold letter indicates a vector.

Equation 10.4 becomes

$$V_1 A_1 = V_2 A_2 \qquad (10.5)$$

where A_1 is the cross-sectional area of the upstream pipe and A_2 is the cross-sectional area of the throat of the flow element.

$$V_1 = V_2 A_2/A_1 \qquad (10.6)$$

By substituting Equation 10.6 in Equation 10.3,

$$V_2 = [2(P_1 - P_2)/\rho]^{1/2}/[1 - (A_2/A_1)^2]^{1/2} \qquad (10.7)$$

85

$$\dot{m} = \rho A_2 V_2 = A_2 [2\rho(P_1 - P_2)]^{1/2}/[1 - (A_2/A_1)^2]^{1/2} \qquad (10.8)$$

where $\dot{m}$ is the mass flow rate.

Since $A_2 = \pi d_2^2/4$, where d_2 is the diameter of the throat of the flow element,

$$\dot{m} = (\pi d_2^2/4)[2\rho(P_1 - P_2)]^{1/2}/[1 - (A_2/A_1)^2]^{1/2} \qquad (10.9)$$

$$\dot{m} = (\pi/8^{1/2})\{[d^{24}\rho(P_1 - P_2)]^{1/2}/[1 - (A_2/A_1)^2]\}^{1/2} \qquad (10.10)$$

where d_2, ρ, V_2, P_1, and P_2 are in consistent units. In SI units, d_2 is in meters (m), ρ is in kg m^{-3}, V_2 is in m s^{-1}, and P_1 and P_2 are in pascals Pa (=Nm^{-2}) = kg m^{-1} s^{-2}, and $\dot{m}$ is in kg s^{-1}.

We now convert to "engineering" units, and use the subscripts E and S to indicate "engineering" and SI units, respectively:

$$d_E/d_S = \text{foot/meter} = 3.28084$$

$$\rho_E/\rho_S = (\text{lb/ft}^3)/(\text{kg/m}^3) = 0.0624279$$

$$P_E/P_S = (\text{lb/ft}^2)/\text{Pa} = 0.671969$$

$$\dot{m}_E/\dot{m}_S = (\text{lb/s})/(\text{kg/s}) = 2.20462$$

Then,

$$(d_E/d_S)^2 [(\rho_E/\rho_S)(P_E/P_S)]^{1/2} = 10.7639 [(0.0624279)(0.671969)]^{1/2} = 2.20462$$

which confirms the ratio of mass flow rates above.

As expected, this ratio is equal to the ratio of the units for the mass flow rate.

NOZZLE DISCHARGE COEFFICIENT

The discharge coefficient, C_D, is the ratio of the actual mass flow rate, $\dot{m}_{act.}$, to the theoretical mass flow rate, $\dot{m}_{th.}$.

$$C_D = \dot{m}_{act.}/\dot{m}_{th.} \qquad (10.11)$$

The actual flow rate through a nozzle is established or measured, using a bell prover and timer, for example. The theoretical flow rate is computed using

$$\dot{m}_{th.} = AC^*P_0/(R_E T_0)^{1/2} \qquad (10.12)$$

where A is the cross-sectional area of the nozzle throat, C^* is the real gas critical flow factor, P_0 is the stagnation pressure at the upstream pressure tap, R_E is the gas constant for air, and T_0 is the stagnation temperature at the nozzle inlet. Thus,

$$C_D = \dot{m}_{act.}(R_E T_0)^{1/2}/AC^*P_0 = (\dot{m}_{act.}R_E^{1/2}/AC^*)(T_0^{1/2}/P_0) \qquad (10.13)$$

DERIVATION OF EXPRESSION FOR C_D

Consider now a gas flowing from a section of piping of cross-sectional area A_1 with velocity V_1 and density ρ_1 through a sonic nozzle into a calibrated bell prover at such a rate that a volume v_2 is collected in the prover in a time interval t.

The density of the gas collected in the prover, with a temperature of T_2 and a pressure of P_2, is ρ_2. The actual mass flow rate is conserved, therefore,

$$\dot{m}_{act.} = \rho_1 A_1 V_1 = \rho_2 v_2/t \qquad (10.14)$$

$$C_D = \rho_2 v_2 (R_E T_0)^{1/2}/tAC^*P_0 = (R_E^{1/2}\rho_2 v_2/AC^*t)(T_0^{1/2}/P_0) \qquad (10.15)$$

Using Jones's development of the air density equation,[1]

$$\rho_2 = MP_2/ZRT_2 \qquad (10.16)$$

where M is the apparent molecular weight of the air, R is the *molar* gas constant, R_E (R/M) is the *specific* gas constant for the air, and Z is the compressibilty factor (the departure of the air from ideality is reflected in the departure of Z from 1). In M and Z we accommodate the possibility that the air is not dry, and the concentration of carbon dioxide in the air is treated as a variable.

We introduce the water vapor mixing ratio, r, defined as

$$r = (\text{mass of water vapor/mass of dry air}) = n_w M_w/n_a M_a \qquad (10.17)$$

where n_w and n_a are the number of moles of water vapor and dry air, respectively and M_w and M_a are molecular weights (apparent, for air) of water vapor and dry air, respectively.

We note that, in the absence of sources and sinks of water vapor in the system, r is conserved. An important implication of the conservation of r is that, if relative humidity can be measured anywhere in the system, r can be calculated and applied to the entire system.

Designating M_w/M_a by ϵ,

$$M = M_a(1 + r)/(1 + r/\epsilon) \qquad (10.18)$$

Then,

$$\rho_2 = M_a[(1 + r)/(1 + r/\epsilon)](P_2/ZRT_2) \qquad (10.19)$$

By substituting Equation 10.16 into Equation 10.15,

$$C_D = (R_E^{1/2}v_2/AC^*t)(T_0^{1/2}/P_0)(MP_2/ZRT_2) \qquad (10.20)$$

and substituting R/M for R_E,

$$C_D = (v_2/AC^*t)(T_0^{1/2}/P_0)(M/R)^{1/2}(P_2/ZT_2) \qquad (10.21)$$

The mass flow rate, $\dot{m}_{act.}$, is equal to

$$(v_2/t)(MP_2/ZRT_2) \qquad (10.22)$$

Therefore,

$$C_D = \dot{m}_{act.}\,[(R/M)T_0]^{1/2}/AC^*P_0 \qquad (10.23)$$

Thus, C_D is determined from the mass flow rate, R/M, A, C^*, the inlet stagnation temperature, T_0, and the stagnation pressure at the upstream pressure tap, P_0.

UNCERTAINTIES IN NOZZLE C_D

With C_D in the form of Equation 10.20, we now investigate the uncertainties in C_D due to uncertainties in the various quantities in Equation 10.20. All uncertainties will be expressed as relative uncertainties, $(\Delta C_D/C_D)_x$ where the x's are quantities in Equation 10.20.

First:

$$(\Delta C_D/C_D)_{T_0} = \tfrac{1}{2}(\Delta T_0/T_0)$$

$$(\Delta C_D/C_D)_{P_0} = -(\Delta P_0/P_0)$$

$$(\Delta C_D/C_D)_A = -(\Delta A/A)$$

$$(\Delta C_D/C_D)_t = -(\Delta t/t)$$

$$(\Delta C_D/C_D)_{C^*} = -(\Delta C^*/C^*)$$

$$(\Delta C_D/C_D)_{RE} = \tfrac{1}{2}(\Delta R_E/R_E)$$

$$(\Delta C_D/C_D)_{\rho_2} = (\Delta \rho_2/\rho_2)$$

$$(\Delta C_D/C_D)_{v_2} = (\Delta v_2/v_2)$$

From the definitions of T_0 and P_0,

$$T_0 \equiv T_1 [1 + (\gamma - 1)M_1^2/2]/[1 + F(\gamma - 1)M_1^2/2] \qquad (10.24)$$

where T_1 is the upstream temperature; γ is the ratio of specific heats of the gas; M_1 is the upstream Mach number; and F is the recovery factor for the temperature sensor, a factor by which the indicated temperature is corrected to the true temperature. The Mach number is defined as the ratio of the velocity of the gas to the speed of sound.

$$P_0 \equiv P_1 [1 + (\gamma - 1)M_1^2/2]^{\gamma/(\gamma - 1)} \qquad (10.25)$$

where P_1 is the upstream pressure.

The upstream Mach number, M_1, can be expressed as

$$M_1 = (P_c/P_1)^{1/\gamma} \{[2/(\gamma - 1)][1 - (P_c/P_1)^{(\gamma - 1)/\gamma}]\}^{1/2}/[\beta^{-4} - (P_c/P_1)^{2/\gamma}]^{1/2} \qquad (10.26)$$

where P_c is the critical pressure, the saturation pressure of a gas at the temperature above which the gas cannot be liquefied by the application of pressure alone; (P_c/P_1) is the critical pressure ratio, i.e., the ratio of the critical pressure, P_c, to the upstream pressure, P_1; and $\beta \equiv d/D$, where d is the diameter of the throat of the nozzle and D is the diameter of the pipe.

Letting $(P_c/P_1) \equiv x$

$$M_1^2 = [2/(\gamma - 1)][x^{2/\gamma} - x^{(\gamma + 1)/\gamma}]/(\beta^{-4} - x^{2/\gamma}) \qquad (10.27)$$

By substituting Equation 10.27 for M_1^2 in Equation 10.24,

$$T_0/T_1 = \{1 + [x^{2/\gamma} - x^{(\gamma + 1)/\gamma}]/(\beta^{-4} - x^{2/\gamma})\}/\{1 + F[x^{2/\gamma} - x^{(\gamma + 1)/\gamma}]/(\beta^{-4} - x^{2/\gamma})\}$$

$$= [\beta^{-4} - x^{(\gamma + 1)/\gamma}]/[\beta^{-4} - Fx^{(\gamma + 1)/\gamma} - (1 - F)x^{2/\gamma}] \qquad (10.28)$$

To assess the uncertainty in the stagnation temperature, T_0, we now study the ratio, T_0/T_1. We calculate the values of T_0/T_1 in Equation 10.28 for various values of the parameters γ and F, and a single value of β in the ranges of interest here.

A typical *high* value of β is 0.325 which corresponds to a low value of β^{-4} of 89.633 and to high values of T_0/T_1. F is less than 1, about 0.7, for example. The range of γ, for air, of interest here is 1.40 to 1.50. A typical value of x is 0.528. For $\beta = 0.325$, $\gamma = 1.40$, x = 0.528, and F = 0.7,

$$T_0/T_1 = (89.633 - 0.335)/(89.633 - 0.234 - 0.120) = 1.00021.$$

For $\beta = 0.325$, $\gamma = 1.45$, x = 0.528, and F = 0.7,

$$T_0/T_1 = (89.633 - 0.340)/(89.633 - 0.238 - 0.124) = 1.00025.$$

For $\beta = 0.325$, $\gamma = 1.50$, $x = 0.528$, and $F = 0.7$,

$T_0/T_1 = (89.633 - 0.345)/(89.633 - 0.242 - 0.128) = 1.00028$.

For $\beta = 0.325$, $\gamma = 1.485$ ($1.50 - 1\%$ of 1.50), $x = 0.528$, and $F = 0.7$,

$T_0/T_1 = (89.633 - 0.343)/(89.633 - 0.240 - 0.127) = 1.00027$.

For $\beta = 0.325$, $\gamma = 1.40$, $x = 0.528$, and $F = 0.1$,

$T_0/T_1 = (89.633 - 0.335)/(89.633 - 0.0335 - 0.361) = 1.00067$.

For $\beta = 0.325$, $\gamma = 1.50$, $x = 0.528$, and $F = 0.1$,

$T_0/T_1 = (89.633 - 0.345)/(89.633 - 0.0345 - 0.384) = 1.00082$.

For all of these examples, the ratio of T_0 to T_1 is very near 1, not greater than 1.00082. The uncertainty in T_0, ΔT_0, is not greater than $1.00082\ \Delta T_1$. If we estimate the uncertainty in T_1 to be 0.1 K or 0.1°C, then the uncertainty in T_0 is not greater than 0.100082 K or 0.100082°C. When estimating uncertainties, two digits are sufficient. Thus, the uncertainty in T_0 is estimated to be 0.10K or 0.10°C.

The relative uncertainties in C_D due to relative uncertainties in P_0, A, t, and v_v are determined by the uncertainties in the measurements of these various quantities. The relative uncertainty in C_D due to the relative uncertainty in ρ_2 is determined by the uncertainties in the measurements of pressure, temperature, and humidity. The relative uncertainty in C_D due to the relative uncertainty in R_E is 0.0018%. The relative uncertainty in C_D due to the relative uncertainty in C^* is approximately 0.073%.

DERIVATION OF DISCHARGE COEFFICIENT, C_D, FOR ORIFICE PLATES

Euler's equation representing the balance of forces in the motion of a frictionless fluid is[2]

$$\rho\mathbf{a} = \rho\mathbf{g} - \nabla P \qquad (10.29)$$

where ρ is the density of the fluid (kg m^{-3}), $\mathbf{a}$ is an arbitrary acceleration (m s^{-2}), $\mathbf{g}$ is the acceleration due to gravity (m s^{-2}), and P is pressure (N m$^{-2} \equiv$ Pa). The bold letters represent vector quantities.

A total differential equation, derived from Euler's equation for steady motion along a streamline[3] is

$$(dV^2)/2 + dP/\rho + g\ dZ = 0 \qquad (10.30)$$

where V is velocity (m s^{-1}) and Z is the vertical coordinate (m).

Integrating Equation 10.30 along a streamline results in Bernoulli's equation:

$$V^2/2 + \int (dP/\rho) + gz = constant \qquad (10.31)$$

For incompressible flow, ρ is constant and Equation 10.31 becomes

$$V^2/2 + P/\rho + gz = constant \qquad (10.32)$$

When we divide Equation 10.32 by g, it becomes

$$(V^2/2g) + (P/\rho g) + z = constant \qquad (10.33)$$

For horizontal flow ($z = z_0$) from a region of cross section A to a region of cross section A_0,

$$(V^2/2g) + (P/\rho g) = (V_0^2/2g) + (P_0/\rho g) \qquad (10.34)$$

Using the equation of *mass continuity*:

$$\rho AV = \rho A_0 V_0, \; AV = A_0 V_0 \qquad (10.35)$$

$$(V_0^2/2g)(A_0/A)^2 + (P/\rho g) = (V_0^2/2g) + (P_0/\rho g) \qquad (10.36)$$

$$V_0^2 = [2(P_0 - P)/\rho]/[(A_0/A)^2 - 1] \qquad (10.37)$$

$$V_0 = [2(P_0 - P)/\rho]^{1/2}/[A_0/A)^2 - 1]^{1/2} \qquad (10.38)$$

The mass flowrate, $\dot{m}$ (kg s^{-1}), through the region of cross section A_0 is

$$\dot{m} = \rho A_0 V_0 = \rho A_0 [2(P_0 - P)/\rho]^{1/2}/[(A_0/A)^2 - 1]^{1/2} \qquad (10.39)$$

In terms of diameters, d_0 and d corresponding to A_0 and A, Equation 10.39 becomes, for circular cross sections:

$$\dot{m} = (\pi \rho d_0^2 V_0/4) = (\pi \rho d_0^2/4) \, [2(P_0 - P)/\rho]^{1/2}/[(d_0/d)^4 - 1]^{1/2} \qquad (10.40)$$

$$\dot{m} = (\pi d_0^2/4) \, [2\rho(P_0 - P)]^{1/2}/[(d_0/d)^4 - 1]^{1/2} \qquad (10.41)$$

The units on either side of Equation 10.41 are

$$(kg \; s^{-1}) = (m^2) \, [(kg \; m^{-3})(Pa)]^{1/2}$$

$$= (m^2) \, [(kg \; m^{-3})(kg \; m^{-1} \; s^{-2})]^{1/2}$$

$$(kg\ s^{-1}) = (m^2)\ [(kg\ m^{-2}\ s^{-1}) = (kg\ s^{-1})$$

Equation 10.41 is an ideal equation that would apply if all of the assumptions on which Equation 10.41 was derived were valid. In practical cases, Equation 10.41 becomes

$$\dot{m} = (\pi d_0^2/4)\ C_D\ [2\rho P_0 - P)]^{1/2}/[(d_0/d)^4 - 1]^{1/2} \qquad (10.42)$$

where C_D is the *discharge coefficient*; it is dimensionless. C_D is determined experimentally and can be considered to vary with temperature, t, and $\dot{m}$ or Reynolds number, Re, that is, $C_D(t, \dot{m})$ or $C_D(t, Re)$.

In the case of flow through an orifice of diameter d_0, the Reynolds number is defined as

$$Re \equiv [V_0 d_0/(\mu_0/\rho_0)] = \rho_0 V_0 d_0/\mu_0 \qquad (10.43)$$

where μ_0 is the viscosity of the fluid.

The mass flow, $\dot{m}$, through the orifice is

$$\dot{m} = \rho_0 V_0 A_0 = \pi \rho_0 V_0 d_0^2/4 \qquad (10.44)$$

By the substitution of Equation 10.44 in Equation 10.43,

$$Re = (4\ \dot{m}/\pi d_0 \mu_0) \qquad (10.45)$$

If $\dot{m}$ is inferred from the volume of water (and its density) collected in a tank during a measured time interval,

$$\dot{m} = v_{tank}\rho_{tank}/(t + \tau) \qquad (10.46)$$

where v_{tank} is the volume of water collected in the tank (m^3), ρ_{tank} is the density of water in the tank (kg m^{-3}), t is the measured collection time, and τ is the diverter correction (s) (see Chapter 18).

Because, generally, the water collected in the tank will not be air-free, the density of the water will be calculated from temperature measurement in the tank using the equation for the density of air-saturated water (see Chapter 3). The density of the water passing through the orifice can be inferred from the calculated density.

By combining Equation 10.46 with Equation 10.42,

$$\dot{m} = v_{tank}\rho_{tank}/(t + \tau) = (\pi d_0^2/4)\ C_D\ [2\rho(P_0 - P)]^{1/2}/[(d_0/d)^4 - 1]^{1/2} \quad (10.47)$$

The discharge coefficient is thus:

$$C_D = [4v_{tank}\rho_{tank}][(d_0/d)^4 - 1]^{1/2}/[\pi d_0^2(t + \tau)][2\rho(P_0 - P)]^{1/2} \quad (10.48)$$

We now return to Equation 10.45. Since Re is dimensionless, the expression on the right-hand side of Equation 10.45 is dimensionless. We confirm this by taking the ratio of the units:

$$(4\dot{m}/\pi d_0\mu_0) = (kg\ s^{-1})/(m)(Pa\cdot s) = (kg\ s^{-1})/(m)(kg\ m^{-1}\ s^{-2}\ s)$$

$$= (kg\ s^{-1})/(kg\ s^{-1}) \tag{10.49}$$

noting that the SI unit of viscosity, μ_0, is the $(Pa\ s) = [kg/(m\ s)]$.

For use in determining the dependence of C_D on Re at the "throat" of the orifice, Re can be calculated using Equation 10.45, where $\dot{m}$ is calculated by the use of Equation 10.46 and μ_0 is taken from tables of the viscosity of water.

The effect of the expansion or contraction of the orifice with temperature can be expressed by the factor:

$$[1 + 2\alpha(T - T_r)]$$

where α is the coefficient of linear expansion with temperature of the material of which the orifice plate is constructed; T is the temperature of the orifice plate, in °C; and T_r is a reference temperature in the same units. The effect on C_D can thus be expressed by

$$C_D = C_{D,r}[1 + 2\alpha(T - T_r)] \tag{10.50}$$

where $C_{D,r}$ is the value of C_D at the reference temperature.

REFERENCES

1. **Jones, F. E.,** An air density equation and the transfer of the mass unit, *J. Res. Natl. Bur. Stand.,* 83, 419, 1978.
2. **Eskanazi, S.,** *Principles of Fluid Mechanics,* 2nd ed., Allyn Bacon, Boston, 1966, 165.
3. **Eskanazi, S.,** *Principles of Fluid Mechanics,* 2nd ed., Allyn Bacon, Boston, 1966, 169.

Chapter 11

AUTOMATIC PIPETS

INTRODUCTION

Automatic pipets can be used to dispense known volumes of water in the volume calibration of tanks which can subsequently be used for flow measurement calibration. It is the purpose of this chapter to illustrate the precision with which known volumes of water can be dispensed from automatic pipets into tanks. A sketch of a typical automatic pipet is illustrated in Figure 11.1.

CALIBRATION OF AUTOMATIC PIPETS

In this work, gravimetric techniques[1] were used to calibrate six automatic pipets with volumes in the approximate range 495 to 716 milliliters (mL).[2] One of the pipets, designated No. 31, was supplied by a commercial glassware supplier; the other five pipets, designated Nos. 1, 2, 12, 22, and 32, were fabricated by personnel of a chemical processing plant.

The gravimetric techniques involve two weighings of a flask: one when empty and one when it contains the quantity of water delivered by the automatic pipet. The weighing flask is first weighed empty; the automatic pipet is then filled from the bottom through the stopcock until excess water passes through the automatic overflow tip. The temperature of the water is measured during the filling.

After the pipet has overflowed, the stopcock is closed and rotated 90° to start the flow from the pipet to the weighing flask. After a 10-second drain time following the cessation of the main flow from the pipet, the stopcock is closed. The weighing flask, containing the water delivered by the pipet, is then reweighed.

The mass difference between the two weighings is divided by the density of water[3] at its temperature in the flask to calculate the volume delivered by the pipet. This volume determination is then adjusted to 20°C using the temperature coefficient of expansion of the material from which the automatic pipet is constructed.

RESULTS AND CONCLUSIONS

Six or seven determinations were made of the volume, V_{20}, delivered by each of the six pipets. The subscript 20 indicates that the volume corresponds to 20°C. The results are listed in Table 11.1. Along with the mean volume for each pipet, V_{20m}, the estimate of standard deviation, SD, and the relative SD, RSD, are listed. RSD = SD/V_{20m}.

FIGURE 11.1.   Sketch of automatic pipet. *(From JNMM, April 1989, with permission.)*

TABLE 11.1
Determinations of Volume (mL)
Delivered at 20°C, V_{20}, by Automatic Pipets

	Pipet No.					
	31	**1**	**2**	**12**	**22**	**32**
	495.326	653.068	667.021	512.406	578.097	716.314
	495.378	653.007	666.970	512.336	578.080	716.336
	495.264	653.060	666.958	512.352	578.096	716.267
	495.339	653.038	667.015	512.335	578.086	716.320
	495.333	653.040	666.941	512.334	578.076	716.297
	495.328	653.059	667.015	512.316	578.082	716.317
					578.095	
			Mean			
V_{20}	495.328	653.045	666.987	512.347	578.087	716.309
SD	0.037	0.022	0.035	0.031	0.0086	0.024
RSD	7.4×10^{-5}	3.4×10^{-5}	5.2×10^{-5}	6.1×10^{-5}	1.5×10^{-5}	3.3×10^{-5}

(From JNMM, April 1989, with permission.)

These results demonstrate that the automatic pipets are capable of being used to deliver volumes of water in the approximate range 497 to 716 mL with a relative precision, RSD, ranging from 1.5 ¥ 10^{-5} to 7.4 ¥ 10^{-5}, i.e., from 0.0015% to 0.0074%.

The automatic pipet is very simple in operation and easy to use very precisely. The quantity delivered by the pipet is very reproducible. The only measurement required is the temperature of the water. These same pipets have been used successfully in the volume calibration of tanks with capacities in the tens-of-liters.

REFERENCES

1. **Lembeck, J.,** The Calibration of Small Volumetric Laboratory Glassware, National Bureau of Standards Internal Report 74-461, December 1974.
2. **Jones, F. E.,** On the use of automatic pipets in volume calibration of accountability tanks, *J. Nucl. Mater. Management,* April, 32, 1989.
3. **Jones, F. E. and Harris, G. L.,** ITS-90 density of water formulation for volumetric standards calibration, *J. Res. Natl. Inst. Stand. Technol.,* 97, 335, 1992.

Chapter 12

REFERENCE METHOD
FOR TESTING HYDROMETERS

INTRODUCTION

A hydrometer is an instrument that can be used to determine the density or the specific gravity of liquids used in flow measurements. A typical hydrometer is shown in Figure 12.1. It is constructed of glass with a bulb partially filled with a suitable ballast material. A slender glass tube enclosing the hydrometer scale is attached to the bulb and the entire instrument is hermetically sealed. The principle on which the hydrometer functions is attributed to Archimedes.

The hydrometer floats with the stem partially immersed in a liquid of which the density or specific gravity (in the range for which the hydrometer is designed) is to be determined. At equilibrium, the hydrometer scale is read at the surface of the liquid. Equilibrium of the hydrometer in the liquid is governed by Archimedes's principle that the upward force on a body immersed in a fluid is equal to the weight (mass × acceleration due to gravity) of the volume of fluid displaced by the body.

This buoyant force, F_b, is expressed by the equation:

$$F_b = \rho v g \tag{12.1}$$

where ρ is density of the fluid; v is the volume of fluid displaced by the body, which is equal to the volume of that part of the hydrometer that is below the surface of the liquid; and g is the local acceleration due to gravity. This equation is to be corrected for the effect of the buoyancy of air[1,2] or vapor surrounding the stem of the hydrometer. The equation is applied to the measurement of the densities of liquids.

There are two classes of hydrometers:

1. Working hydrometers, used in science and industry for liquid density measurements of moderate accuracy and precision.
2. Reference standard hydrometers, used exclusively for the calibration of working hydrometers.

Reference standard hydrometers can be calibrated by relating their flotation levels (indicated on the hydrometer scales) to liquid bath densities independently determined by hydrostatic weighing of solid object density standards.[3]

Working hydrometers are tested or verified by floating the working hydrometer together with a reference standard hydrometer in a suitable liquid in a comparator. The readings of the two hydrometers are compared, and correction to the reading for the working hydrometer is determined.

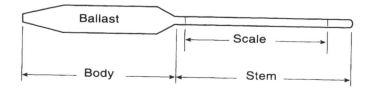

FIGURE 12.1. A typical hydrometer. *(From ASTM Standard:E 100-81. Copyright ASTM. Reprinted with permission.)*

HYDROMETER CALIBRATION METHOD

The density of a liquid can be inferred from the differential pressure measured between two probes immersed at different heights in the liquid. The method for hydrometer calibration discussed here is based on this old principle.

The relationship between the differential pressure (DP) between two points (or levels) in a liquid, separated vertically by a distance h, and the density of the liquid is

$$DP = \rho g h \tag{12.2}$$

where ρ is the density of the liquid and g is the local acceleration due to gravity. The ratio of DP to ρ is a measure of the product (gh). Therefore, a series of measurements of DP in a liquid of known density provides a determination of (gh) with an estimate of precision from the multiple measurements. The value of (gh) thus determined then becomes a calibration factor relating density to DP.

In the practical or experimental determination of DP, a number of parameters must be considered or corrected for. In the experimental work described in this chapter, the DP between two levels in the liquid is determined by the use of two "bubbler tubes" vertically separated by a distance h. Air is forced through the bubbler tubes and bubbles form and break off the ends of the tubes. The pressure at the effective tip of each tube is transmitted to one side of a pressure-measuring device by the air.

The pressure difference *at the pressure-measuring device*, designated here $(P_{GH} - P_{GO})$, is measured by the pressure-measuring device. In the present work, the pressure-measuring device is an electromanometer with a Bourdon-type pressure-sensing element.

The use of bubbler tubes in the determination of pressure in liquids also is an old method that was patented in the U.S. as early as 1923.

The DP *of interest*, designated here $(P_b - P_u)$, is the difference in pressure at the effective tips of the bubbler tubes (b indicating the lower tip and u indicating the upper tip).

In Figure 12.2, a typical vessel containing some liquid and the measurement system are sketched. The various designations on the sketch are defined as follows:[4]

P_{GH} — pressure at the high-pressure side of the electromanometer.

P_{GO} — pressure at the low-pressure side of the electromanometer.

P_b — pressure at the level of the tip of the longer bubbler tube.

P_u — pressure at the level of the tip of the shorter bubbler tube.

P_{t1} — pressure associated with the formation and the surface tension of the bubble on the tip of the longer bubbler tube.

P_{Dh} — pressure drop in the longer bubbler tube due to air flow through the tube.

$(\rho_a)_h$ — density of air in the longer bubbler tube.

g — local acceleration due to gravity.

h — vertical length of the longer bubbler tube.

P_{t2} — pressure associated with the formation and the surface tension of the bubble on the tip of the shorter bubbler tube.

P_{Ds} — pressure drop in the shorter bubbler tube due to air flow through the tube.

$(\rho_a)_s$ — density of air in the shorter bubbler tube.

s — vertical length of the shorter bubbler tube.

ρ — density of the liquid.

L_h — height of the liquid surface above the tip of the longer bubbler tube.

L_s — height of the liquid surface above the tip of the shorter bubbler tube.

$(\rho_a)_p$ — density of air in the space above the liquid surface.

P_f — pressure at the liquid surface.

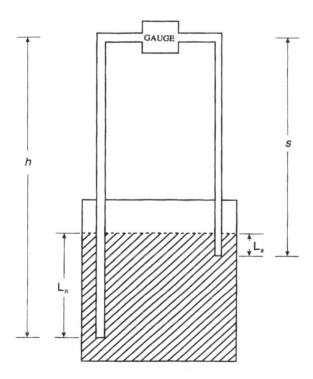

FIGURE 12.2. A typical vessel containing liquid and instrumentation.

The following equations relate the parameters listed above:

$$P_{GH} = P_b + P_{t1} + P_{Dh} - (\rho_a)_h gh \qquad (12.3)$$

$$P_{GO} = P_u + P_{t2} + P_{Ds} - (\rho_a)_s gs \qquad (12.4)$$

$$P_b = \rho_w g L_h + P_f \qquad (12.5)$$

$$P_u = \rho_w g L_s + P_f \qquad (12.6)$$

$$P_{GH} - P_{GO} = (P_b - P_u) + (P_{t1} - P_{t2}) + (P_{Dh} - P_{Ds})$$

$$- g[(\rho_a)_h h - (\rho_a)_s s] \qquad (12.7)$$

For tubing of sufficiently large inside diameter, d, the difference in pressure drop in the two tubes due to the flow of air through them, $(P_{Dh} - P_{Ds})$, can be made negligible since the pressure drop in the tubing varies inversely with d^{4}.[5]

The pressure associated with the formation and the surface tension of the bubble on the tip of a bubbler tube, P_{t1} or P_{t2}, is given by[6,7]

$$P_{t1} = [\rho_w - (\rho_a)_h] g l_1 + 2\gamma_1/R_1 \qquad (12.8)$$

and

$$P_{t2} = [\rho_w - (\rho_a)_s] g l_2 + 2\gamma_2/R_2 \qquad (12.9)$$

where l is the vertical length of the bubble on the tip, γ is the surface tension for air and the liquid, and R is the radius of curvature of the bubble at its lowest point.

It can be assumed[7] that in the present case $l_1 = l_2$ and $R_1 = R_2$. Also, $\gamma_1 = \gamma_2$. It can be shown that $(P_{t1} - P_{t2})$ is smaller than $\{-g[(\rho_a)_h h - (\rho_a)_s s]\}$ by more than 2 orders of magnitude and can be ignored. Also,

$$(P_b - P_u) = \rho_w g(L_h - L_s) \qquad (12.10)$$

After setting $(P_{Dh} - P_{Ds})$ effectively equal to zero, Equation 12.7 becomes

$$(P_{GH} - P_{GO}) = \rho_w g(L_h - L_s) - g[(\rho_a)_h h - (\rho_a)_s s] \qquad (12.11)$$

and

$$(P_{GH} - P_{GO}) + g[(\rho_a)_h h - (\rho_a)_s s] = \rho_w g(L_h - L_s) \qquad (12.12)$$

$(L_h - L_s)$ is seen to be the vertical distance between the two bubbler tube tips. Thus, Equation 12.11 is of the form of Equation 12.2.

A set of measurements of $(P_{GH} - P_{GO})$ for known density of water (ρ_w) as the calibrating liquid and calculations of the quantities in the second term (designated δP) on the left side of Equation 12.12 provides determinations of $[g(L_h - L_s)]$, that is,

$$[g(L_h - L_s)] = \{(P_{GH} - P_{GO}) + g[(\rho_a)_h h - (\rho_a)_s s]\}/\rho_w \qquad (12.13)$$

After $[g(L_h - L_s)]$, which can be considered to be a calibration constant, has been determined by water calibration, Equation 12.13 can be rearranged to be used to determine the density, ρ_q, of another liquid:

$$\rho_q = \{(P_{GH} - P_{GO}) + g[(\rho_a)_h h - (\rho_a)_s s]\}/[g(L_h - L_s)] \qquad (12.14)$$

It should be emphasized that g or L_h and L_s need not be known; the product $[g(L_h - L_s)]$ is determined directly and used subsequently in the determination of the densities of other liquids.

DENSITY OF WATER

Because the relationship between the density of water and its temperature is well known, it is the ideal liquid for making the determination of $[g(L_h - L_s)]$ for use in Equation 12.13. In Chapter 3, new formulations for the density of air-free and air-saturated water as functions of temperature on the 1990 International Temperature Scale (ITS-90)[8] and tables of calculated values[8] were presented. The formulation for air-free water was used to calculate values of air-free water density, ρ_w, for insertion in Equation 12.13 for the determination of $[g(L_h - L_s)]$, the calibration constant in Equation 12.14.

The calculation equation[8] for temperature in the range 5 to 40°C and for ambient pressure in the vicinity of 101.325 kPa (1 atmosphere) is

$$\rho_w = 999.85308 + 6.32693 \times 10^{-2} \, t - 8.523829 \times 10^{-3} \, t^2$$

$$+ 6.943248 \times 10^{-5} \, t^3 - 3.821216 \times 10^{-7} \, t^4 \qquad (12.15)$$

where ρ_w is in kg m^{-3} and t is temperature in °C on ITS-90.

Equation 12.15 applies to freshly distilled water. However, in the temperature range of interest, 15 to 25°C, if the water is saturated with air, air saturation reduces the water density by only 2 or 3 parts per million, which can be neglected for present purposes.

EXPERIMENTAL

A differential pressure electromanometer (with accuracy of ±0.003% of full scale and medium-term stability of ±0.005% of reading) was connected across two vertically separated bubbler tube probes, the tips of both of which were immersed in a liquid in a chemical process vessel. The liquid was either water or a 2% nitric acid solution. The probes were stainless steel tubes of 0.019-m inside diameter, with a vertical separation of about 0.254 m. A flow of air of 8 mL s^{-1} was maintained through each of the probes, communicating a differential pressure to the electromanometer. The temperature of the liquid was measured.

With the tips of both probes immersed in water and with air flowing through the probes, a series of measurements of differential pressure ($P_{GH} - P_{GO}$) and temperature was made for each of four runs.

The quantity $g[(\rho_a)_h h - (\rho_a)_s s]$, δP, was calculated for each point, using the formulations of References 9 and 10 to calculate air density, and added to each respective value of ($P_{GH} - P_{GO}$); the sum of the two terms is designated ΔP. For these four runs, the value of δP was a constant 3.1 Pa. The data for the four runs are listed in Tables 12.1 through 12.4. The tables list the temperature of the water, in °C on ITS-90; ΔP in Pa; the density of water calculated using Equation 12.15, in kg m^{-3}; and calculated values of $g(L_h - L_s)$, in m^2 s^{-2}, reduced to values at 20°C using the equation:

$$[g(L_h - L_s)]_{20} = [g(L_h - L_s)]_t \, [1 + \alpha(20 - t)] \qquad (12.16)$$

where α, the thermal coefficient of linear expansion of the material of which the probes were constructed, is 15.9×10^{-6} (°C)$^{-1}$ in the present case.

Also included on each table are the mean of the values of $[g(L_h - L_s)]_{20}$; the estimate of the standard deviation, SD; the estimate of the standard deviation of the mean, SD/n$^{1/2}$ (SE); and the relative standard deviation of the mean, (SE)/mean, expressed in %.

An analysis of variance, ANOVA, of the data for the four runs in water was made to test the hypothesis that all of the determinations of $[g(L_h - L_s)]$ are random samples from the same population. The value of F for the ANOVA was 0.33, which is sufficiently small that the hypothesis can be accepted. Since all the determinations (149 in number) can be considered to be random samples from the same population, all of them were used to determine an overall mean value and corresponding values of SD, SE, and relative SE. The overall mean value of $[g(L_h - L_s)]_{20}$ was 2.4726 m^2 s^{-2}, the SD was 0.0028 m^2 s^{-2}, SE was 0.00023 m^2 s^{-2}, and the relative SE was 0.0093%.

Again, g or L_h and L_s need not be known separately; rather, the product $[g(L_h - L_s)]$ is determined from measured ΔP and known ρ_w. However, the value of the effective separation of the probe tips, $L_h - L_s$, can be determined if the acceleration due to gravity at the location at which the measurements are made is known. At the location of the vessel in which the measurements were made, the value of g is known to be 9.79547 m s^{-2}. The value of $L_h - L_s$ is then 0.2524 m, which is near the design value of 0.254 m. In the use of the system to determine the densities of other liquids, nitric acid solutions in the present case, the overall mean value of $[g(L_h - L_s)]$ would be used.

Because the value of $[g(L_h - L_s)]$ determined with water corresponds to a reference temperature of 20°C, a correction for thermal expansion or contrac-

| | **TABLE 12.1** | | | | | **TABLE 12.2** | | |
| | **Data for Run No. 1 in Water** | | | | | **Data for Run No. 2 in Water** | | |

Temperature (°C)	ΔP (Pa)	ρ_w (kg m^{-3})	$[g(L_h - L_s)]_{20}$ (m^2 s^{-2})	Temperature (°C)	ΔP (Pa)	ρ_w (kg m^{-3})	$[g(L_h - L_s)]_{20}$ (m^2 s^{-2})
19.4	2470.7	998.325	2.4748	19.6	2472.8	998.285	2.4770
19.7	2471.4	998.265	2.4757	19.8	2472.1	998.244	2.4764
19.6	2469.0	998.285	2.4732	19.8	2467.6	998.244	2.4719
19.7	2472.5	998.265	2.4768	19.8	2469.7	998.244	2.4740
19.7	2464.5	998.265	2.4688	19.8	2468.3	998.244	2.4726
19.7	2468.3	998.265	2.4726	19.8	2474.5	998.244	2.4789
19.7	2469.0	998.265	2.4733	19.8	2471.8	998.244	2.4761
19.7	2471.1	998.265	2.4654	19.8	2472.1	998.244	2.4764
19.7	2472.8	998.265	2.4771	19.8	2470.0	998.244	2.4743
19.7	2467.3	998.265	2.4716	19.8	2470.7	998.244	2.4750
19.7	2471.4	998.265	2.4757	19.9	2469.0	998.224	2.4734
19.7	2472.1	998.265	2.4764	19.8	2473.8	998.244	2.4782
19.7	2468.3	998.265	2.4726	19.8	2469.0	998.244	2.4733
19.7	2470.4	998.265	2.4747	19.9	2471.8	998.224	2.4762
19.7	2469.3	998.265	2.4736	19.9	2470.0	998.224	2.4744
19.7	2470.0	998.265	2.4743	19.9	2470.4	998.224	2.4748
19.7	2470.4	998.265	2.4747	19.9	2469.0	998.224	2.4734
19.7	2470.0	998.265	2.4743	20.0	2470.4	998.203	2.4748
19.7	2466.3	998.265	2.4706	20.1	2469.7	998.183	2.4742
19.7	2465.9	998.265	2.4702	20.1	2466.6	998.183	2.4711
19.7	2469.3	998.265	2.4736	20.2	2465.6	998.162	2.4701
19.7	2467.6	998.265	2.4719	20.2	2469.7	998.162	2.4742
19.7	2468.0	998.265	2.4723	20.2	2465.6	998.162	2.4701
19.7	2465.6	998.265	2.4699	20.2	2465.6	998.162	2.4701
19.7	2463.5	998.265	2.4678	20.2	2466.6	998.162	2.4771
19.7	2462.8	998.265	2.4671	20.2	2465.9	998.162	2.4704
19.7	2468.0	998.265	2.4723	20.3	2466.9	998.141	2.4715
19.7	2467.6	998.265	2.4719	20.3	2466.9	998.141	2.4715
19.7	2469.7	998.265	2.4740	20.3	2462.5	998.141	2.4671
19.7	2465.9	998.265	2.4702	20.3	2461.8	998.141	2.4664
19.7	2468.0	998.265	2.4723	20.3	2463.8	998.141	2.4684
19.7	2468.3	998.265	2.4726	20.3	2462.5	998.141	2.4671
19.7	2467.6	998.265	2.4719	20.3	2465.6	998.141	2.4702
19.7	2468.3	998.265	2.4726	20.4	2463.8	998.120	2.4684
19.7	2467.3	998.265	2.4716	20.4	2469.0	998.120	2.4737
19.7	2466.9	998.265	2.4712	20.4	2466.6	998.120	2.4712
				20.4	2466.9	998.120	2.4715
				20.4	2469.0	998.120	2.4737
				20.4	2463.8	998.120	2.4684
				20.4	2467.3	998.120	2.4719

Mean $[g(L_h - L_s)]_{20}$ = 2.4725 m^2 s^{-2}
SD = 0.0026 m^2 s^{-2}
SE = 0.00044 m^2 s^{-2}
Relative SE = 0.018%

Mean $[g(L_h - L_s)]_{20}$ = 2.4727 m^2 s^{-2}
SD = 0.0031 m^2 s^{-2}
SE = 0.00050 m^2 s^{-2}
Relative SE = 0.020%

TABLE 12.3
Data for Run No. 3 in Water

Temperature (°C)	ΔP (Pa)	ρ_w (kg m^{-3})	$[g(L_h - L_s)]_{20}$ (m^2 s^{-2})
20.1	2468.7	998.183	2.4732
20.4	2470.7	998.120	2.4754
20.5	2471.1	998.099	2.4758
20.5	2463.5	998.099	2.4682
20.4	2472.5	998.120	2.4772
20.5	2468.7	998.099	2.4734
20.5	2472.5	998.099	2.4772
20.4	2471.8	998.120	2.4765
20.4	2470.4	998.120	2.4751
20.4	2472.1	998.120	2.4768
20.4	2469.7	998.120	2.4744
20.4	2470.4	998.120	2.4751
20.4	2471.4	998.120	2.4761
20.4	2470.0	998.120	2.4747
20.4	2471.8	998.120	2.4765
20.4	2470.7	998.120	2.4754
20.4	2471.1	998.120	2.4758
20.4	2465.6	998.120	2.4702
20.4	2465.9	998.120	2.4705
20.4	2469.4	998.120	2.4741
20.4	2468.3	998.120	2.4729
20.4	2470.0	998.120	2.4747
20.4	2470.0	998.120	2.4747
20.4	2471.1	998.120	2.4758
20.4	2469.7	998.120	2.4744
20.4	2471.4	998.120	2.4761
20.4	2467.3	998.120	2.4719
20.4	2467.3	998.120	2.4719
20.4	2468.3	998.120	2.4729
20.4	2464.2	998.120	2.4688
20.4	2461.4	998.120	2.4660
20.4	2464.5	998.120	2.4701
20.4	2464.9	998.120	2.4695
20.3	2465.1	998.141	2.4697
20.3	2463.8	998.141	2.4684
20.3	2466.3	998.141	2.4709
20.3	2466.9	998.141	2.4715
20.3	2468.0	998.141	2.4726
20.3	2464.5	998.141	2.4691
20.3	2464.9	998.141	2.4695
20.3	2466.6	998.141	2.4712
20.3	2464.5	998.141	2.4691
20.4	2462.1	998.120	2.4667
20.3	2470.0	998.141	2.4746
20.4	2470.4	998.120	2.4751

TABLE 12.3 (CONTINUED)

Mean $[g(L_h - L_s)]_{20}$ = 2.4729 m^2 s^{-2}

SD = 0.0030 m^2 s^{-2}

SE = 0.00045 m^2 s^{-2}

Relative SE = 0.018%

TABLE 12.4
Data for Run No. 4 in Water

Temperature (°C)	ΔP (Pa)	ρ_w (kg m^{-3})	$[g(L_h - L_s)]_{20}$ (m^2 s^{-2})
20.7	2464.9	998.056	2.4697
20.6	2466.6	998.077	2.4714
20.6	2470.0	998.077	2.4748
20.6	2468.0	998.077	2.4728
20.4	2469.7	998.120	2.4744
20.8	2470.4	998.035	2.4753
20.8	2470.0	998.035	2.4749
20.9	2468.7	998.013	2.4736
20.9	2466.9	998.013	2.4718
20.8	2466.6	998.035	2.4715
20.8	2470.0	998.035	2.4749
20.8	2468.3	998.035	2.4732
20.8	2469.7	998.035	2.4746
20.9	2469.0	998.013	2.4739
20.9	2463.2	998.013	2.4681
20.9	2464.5	998.013	2.4694
20.9	2469.0	998.013	2.4739
20.9	2464.5	998.013	2.4694
20.9	2466.9	998.013	2.4718
20.9	2464.5	998.013	2.4694
20.9	2466.9	998.013	2.4718
21.0	2464.2	997.991	2.4692
21.0	2463.8	997.991	2.4688
21.0	2467.3	997.991	2.4723
21.0	2467.6	997.991	2.4726
21.0	2468.3	997.991	2.4733
21.0	2464.9	997.991	2.4699
21.0	2467.6	997.991	2.4726

Mean $[g(L_h - L_s)]_{20}$ = 2.4721 m^2 s^{-2}

SD = 0.0022 m^2 s^{-2}

SE = 0.00044 m^2 s^{-2}

Relative SE = 0.0017%

tion of the probes would be made when the system is at other temperatures. The corrected value at temperature t, $[g(L_h - L_s)]_t$, is given by

$$[g(L_h - L_s)]_t = [g(L_h - L_s)]_{20} [1 + \alpha(t - 20)] \qquad (12.17)$$

where, as in Equation 12.16, α is the thermal coefficient of linear expansion for the material of which the probes are constructed.

Measurements in Nitric Acid

In the same vessel in which the system was calibrated, the probes were immersed in 2% nitric acid solutions. Thus, the mean value of $[g(L_h - L_s)]_{20}$ determined above was used to determine the density of the acid solutions, using Equation 12.14. $[g(L_h - L_s)]_{20}$ was temperature corrected using Equation 12.17.

Four runs were made with acid solutions in the vessel. The data for the four runs are listed in Tables 12.5 through 12.8. For these four runs, the value of δP was a constant 3.1 Pa. The tables list the temperature of the solution, in °C; ΔP, in Pa; $[g(L_h - L_s)]_t$, in m^2 s^{-2}; the calculated density of the solution, ρ_t, at the temperature of the solution, in kg m^{-3}; and ρ_{20}, the calculated density of the solution reduced to the reference temperature, 20°C, in kg m^{-3}. Also included in each table is the mean ρ_{20}, SD, SE, and the relative SE.

An analysis of variance, ANOVA, of the data for the four acid solution runs indicated that not all of the determinations of mean ρ_q were random samples from the same population, that is, there were significant run-to-run differences. An obvious explanation is that the concentration of the acid solution (and, therefore, the density) was not the same for each run. The mean values of calculated acid solution density at the reference temperature were 1009.80, 1009.95, 1010.26, and 1020.19 kg m^{-3}. The corresponding values of relative SE are 0.011, 0.012, 0.012, and 0.0077%.

APPLICATION OF THE METHOD
TO HYDROMETER CALIBRATION

In hydrometer calibration, the hydrometer to be calibrated, verified, or tested would be floated in the test liquid in a comparator vessel and the temperature of the liquid would be measured. Two probes, with a vertical separation of $L_h - L_s$, connected across the differential pressure measuring device would be immersed in the liquid.

Initially, the mean $[g(L_h - L_s)]$ for the probes would be determined by immersing the probes in water. The density of the water would be calculated for measured temperature using Equation 12.15. The water would be stirred and multiple determinations of ΔP and multiple measurements of temperature would be made. The value of $[g(L_h - L_s)]$ would then be determined from the ratio of ΔP to ρ_w. The mean value of $[g(L_h - L_s)]_{20}$ would be a calibration factor

TABLE 12.5
Data for Run No. 1 in Acid Solution

Temperature ($°C$)	ΔP (Pa)	$[g(L_h - L_s)]$ ($m^2 s^{-2}$)	ρ_t ($kg\ m^{-3}$)	ρ_{20} ($kg\ m^{-3}$)
22.24	2494.5	2.4727	1008.8	1009.3
22.43	2495.9	2.4727	1009.4	1009.9
22.45	2496.9	2.4727	1009.8	1010.4
22.47	2498.0	2.4727	1010.2	1010.8
22.49	2496.9	2.4727	1009.8	1010.4
22.55	2498.3	2.4727	1010.4	1011.0
22.56	2498.0	2.4727	1010.2	1010.8
22.56	2496.6	2.4727	1009.7	1010.3
22.56	2494.5	2.4727	1008.8	1009.4
22.56	2495.9	2.4727	1009.4	1010.0
22.54	2496.2	2.4727	1009.5	1010.1
20.74	2497.6	2.4726	1010.1	1010.3
20.77	2494.9	2.4726	1009.0	1009.2
20.77	2496.6	2.4726	1009.7	1009.9
20.75	2493.8	2.4726	1008.6	1008.8
20.75	2496.9	2.4726	1009.8	1010.2
20.68	2495.2	2.4726	1009.1	1009.3
20.68	2494.9	2.4726	1009.0	1009.2
20.67	2497.6	2.4726	1010.1	1010.3
20.66	2495.2	2.4726	1009.1	1009.2
20.64	2496.6	2.4726	1009.7	1009.8
20.62	2495.6	2.4726	1009.3	1009.4
20.61	2497.6	2.4726	1010.1	1010.2
20.56	2495.9	2.4726	1009.4	1009.5
20.55	2495.2	2.4726	1009.1	1009.2
20.54	2495.9	2.4726	1009.4	1009.5
20.53	2494.5	2.4726	1008.9	1009.0
20.51	2496.6	2.4726	1009.7	1009.8
20.50	2494.5	2.4726	1008.9	1009.0
20.49	2496.9	2.4726	1009.8	1009.9

Mean ρ_{20} = 1009.80 kg m^{-3}
SD = 0.59 kg m^{-3}
SE = 0.11 kg m^{-3}
Relative SE = 0.011%

for the system and would be used subsequently to determine the density of test liquids in the comparator.

The water would then be removed and a liquid with nominal density in the desired range for the calibration, verification, or testing of a hydrometer would be introduced into the comparator vessel. The hydrometer would be floated in the liquid alongside the bubbler tubes. A series of determinations of ΔP and measurements of temperature would be made and the density of

TABLE 12.6
Data for Run No. in Acid Solution

Temperature (°C)	ΔP (Pa)	$[g(L_h - L_s)]$ ($m^2 s^{-2}$)	ρ_t (kg m^{-3})	ρ_{20} (kg m^{-3})
20.34	2499.0	2.4726	1010.7	1010.8
20.34	2499.0	2.4726	1010.7	1010.8
20.34	2499.3	2.4726	1010.8	1010.9
20.34	2499.7	2.4726	1011.0	1011.1
20.34	2499.3	2.4726	1010.8	1010.9
20.35	2497.6	2.4726	1010.1	1010.2
20.37	2498.0	2.4726	1010.3	1010.4
20.36	2499.9	2.4726	1010.7	1010.8
20.37	2500.0	2.4726	1011.1	1011.2
20.36	2495.9	2.4726	1009.4	1009.5
20.37	2495.9	2.4726	1009.4	1009.5
20.38	2496.6	2.4726	1009.7	1009.8
20.40	2494.2	2.4726	1008.7	1008.8
20.41	2495.9	2.4726	1009.4	1009.5
20.41	2495.6	2.4726	1009.3	1009.4
20.42	2495.9	2.4726	1009.4	1009.5
20.43	2495.9	2.4726	1009.4	1009.5
20.44	2495.6	2.4726	1009.3	1009.4
20.44	2495.9	2.4726	1009.4	1009.5
20.44	2495.9	2.4726	1009.4	1009.5
20.44	2497.3	2.4726	1010.0	1010.1
20.44	2496.6	2.4726	1009.7	1009.8
20.44	2494.5	2.4726	1008.9	1009.0
20.45	2496.2	2.4726	1009.5	1009.6
20.45	2495.9	2.4726	1009.4	1009.5
20.47	2497.3	2.4726	1010.0	1010.1
20.47	2496.2	2.4726	1009.5	1009.6
20.47	2496.9	2.4726	1009.8	1009.9

Mean ρ_{20} = 1009.95 kg m^{-3}
SD = 0.66 kg m^{-3}
SE = 0.13 kg m^{-3}
Relative SE = 0.012%

the liquid would be calculated from the ratios of ΔP to $[g(L_h - L_s)]_t$. The values of specific gravity would be calculated from the density values. The usual observations would be made on the working hydrometer and compared with the density or specific gravity of the liquid. For each liquid, for most reliable results, a water calibration would precede the determination of the density of the liquid.

A reference standard hydrometer could also be immersed in the liquid along with the bubbler tubes and hydrometer measurements could be made between determinations of density.

TABLE 12.7
Data for Run No. 3 in Acid Solution

Temperature (°C)	ΔP (Pa)	$[g(L_h - L_s)]$ $(m^2\ s^{-2})$	ρ_t (kg m⁻³)	ρ_{20} (kg m⁻³)
20.66	2500.4	2.4726	1011.2	1011.3
20.66	2496.9	2.4726	1009.8	1009.9
20.59	2498.0	2.4726	1010.3	1010.4
20.58	2497.6	2.4726	1010.1	1010.2
20.56	2496.7	2.4726	1009.7	1009.8
20.55	2499.7	2.4726	1011.0	1011.1
20.53	2499.0	2.4726	1010.7	1011.8
20.52	2498.7	2.4726	1010.6	1010.7
20.51	2499.3	2.4726	1010.8	1010.9
20.52	2496.1	2.4726	1009.5	1009.6
20.54	2499.7	2.4726	1011.0	1011.1
20.53	2499.0	2.4726	1010.7	1010.8
20.53	2496.9	2.4726	1009.8	1009.9
20.53	2499.7	2.4726	1011.0	1011.1
20.52	2496.2	2.4726	1009.5	1009.6
20.53	2496.2	2.4726	1009.5	1009.6
20.53	2496.6	2.4726	1009.7	1009.8
20.54	2496.2	2.4726	1009.5	1009.6
20.56	2496.9	2.4726	1009.8	1009.9
20.58	2496.9	2.4726	1009.8	1009.9
20.62	2496.6	2.4726	1009.7	1009.8
20.63	2498.0	2.4726	1010.3	1010.4
20.65	2496.9	2.4726	1009.8	1009.9
20.66	2497.3	2.4726	1010.0	1010.1
20.67	2496.9	2.4726	1009.8	1009.9
20.67	2496.9	2.4726	1009.8	1009.9
20.75	2497.3	2.4726	1010.0	1010.1

Mean ρ_{20} = 1010.26 kg m⁻³
SD = 0.62 kg m⁻³
SE = 0.12 kg m⁻³
Relative SE = 0.012%

SUMMARY

A system for the determination of the density of liquids for the calibration, verification, or testing of hydrometers has been described. The system is based on sound scientific principles, is simple, and is easy to use. It is easily calibrated using pure distilled water. Recalibration is very rapid. The precision of the system, as demonstrated by results for water and for nitric acid solutions, is approximately 1 part in 10,000 or 0.01%. This precision is ideal for the testing of hydrometers the precision of which is about 1 part in 1,000 or 0.1%.

The system is a very precise, simple, easy-to-use densitometer system that can be used to determine the densities of liquids. In the experimental phase of

TABLE 12.8
Data for Run No. 4 in Acid Solution

Temperature (°C)	ΔP (Pa)	$[g(L_h - L_s)]$ (m² s⁻²)	ρ_t (kg m⁻³)	ρ_{20} (kg m⁻³)
21.19	2498.3	2.4726	1010.4	1010.7
21.16	2499.3	2.4726	1010.8	1011.1
21.16	2498.3	2.4726	1010.4	1010.7
21.10	2498.3	2.4726	1010.4	1010.6
21.10	2498.7	2.4726	1010.6	1010.8
21.25	2498.3	2.4726	1010.4	1010.7
21.29	2497.6	2.4727	1010.1	1010.4
21.35	2498.0	2.4727	1010.2	1010.5
21.33	2497.6	2.4727	1010.1	1010.4
21.33	2496.9	2.4727	1009.8	1010.1
21.35	2496.6	2.4727	1009.7	1010.0
21.37	2497.3	2.4727	1009.9	1010.2
21.70	2495.9	2.4727	1009.4	1009.8
21.72	2595.9	2.4727	1009.4	1009.8
21.72	2495.9	2.4727	1009.4	1009.8
21.74	2496.2	2.4727	1009.5	1009.9
21.75	2496.9	2.4727	1009.8	1010.2
21.76	2496.2	2.4727	1009.5	1009.9
21.72	2496.2	2.4727	1009.5	1009.9
21.70	2496.2	2.4727	1009.5	1009.9
21.74	2496.6	2.4727	1009.7	1010.1
21.77	2496.2	2.4727	1009.5	1009.9
21.77	2495.2	2.4727	1009.1	1009.5
21.77	2496.2	2.4727	1009.5	1009.9
21.74	2496.2	2.4727	1009.5	1009.9
21.73	2496.9	2.4727	1009.8	1010.2

Mean ρ_{20} = 1010.19 kg m⁻³
SD = 0.40 kg m⁻³
SE = 0.078 kg m⁻³
Relative SE = 0.0077%

the work reported here, the system was not optimized. With optimization, the performance of the system could be even better.

REFERENCES

1. **Schoonover, R. M. and Jones, F. E.,** Air density correction in high-accuracy weighing on analytical balances, *Anal. Chem.*, 53, 900, 1981.
2. **Jones, F. E.,** The air density equation and the transfer of the mass unit, *J. Res. Natl. Bur. Stand.*, 83, 419, 1978.

3. **Bowman, H. A. and Schoonover, R. M.,** Procedure for high precision density determination by hydrostatic weighing, *J. Res. Natl. Bur. Stand.*, 71C, 179, 1967.

4. **Jones, F. E.,** A tank volume calibration algorithm, *Nucl. Mater. Management*, Spring 1984, 16.

5. **Poiseuille, J. L. M.,** Experimental investigations upon the flow of liquids in tubes of very small diameter, *Compt. Rend.*, 1041, 1840.

6. **Landau, L. D. and Lifshitz, E. M.,** *Fluid Mechanics,* Pergamon Press, London, 1959, 231.

7. **Gaigalas, A. K. and Robertson, B.,** Time dependence of pressure in a bubbler tube, *AIChe J.*, 28, 922, 1982, 922.

8. **Jones, F. E. and Harris, G. L.,** ITS-90 density of water formulation for volumetric standards calibration, *J. Res. Natl. Inst. Stand. Technol.*, 97, 335, 1992.

9. **Giacomo, P.,** Equation for the determination of density of moist air, *Metrologia*, 18, 33, 1982.

10. **Davis, R. S.,** Equation for the determination of density of moist air (1981-1991), *Metrologia*, 29, 67, 1992.

Chapter 13

TREATMENT OF LAMINAR FLOWMETER
CALIBRATION DATA

INTRODUCTION

Dried air or nitrogen is used as the flowing gas in the calibration of laminar flowmeters. Subsequently, the calibrated flowmeters are often used to measure the flow of other gases. Therefore, the properties of the working gas (and of the calibration gas) must be considered if flow measurements are to be accurate.

In this chapter, a corrected form of the Poiseuille equation is developed and a set of calibration data for dried air, nitrogen, helium, and argon is used to determine the values of the coefficients in the equation. The resulting single equation represents the data for the four gases over the range of calibration, and it would presumably be applicable for other well-behaved gases over the same range.

DEVELOPMENT OF THE EQUATION

The Poiseuille equation[1,2] relating the volume flow rate, Q, of a gas of viscosity, μ', through a laminar flow element can be written in the general form:

$$Q = B'(\Delta P/\mu')l \qquad (13.1)$$

where ΔP is pressure drop across the element, B' is a constant that includes the linear dimensions of a transverse section through the element, and l is the length of the element. The equation is applicable to laminar flow in narrow elements. However, certain corrections must be made.[3]

A correction to account for the kinetic energy of fluid leaving a tube was made by Hagenbach.[4] He corrected the head h (h = $\Delta P/\rho g$, where ρ is the density of the fluid and g is the acceleration due to gravity) by subtracting the quantity $0.7937\ V^2/g$, where V is the velocity of the flowing fluid.

Reynolds[5] subtracted $0.7525\ V^2/g$ and Wilberforce[6] subtracted V^2/g.

Fain[7] considered the "inertial" pressure loss resulting from acceleration and deceleration at the entrance and exit of a laminar flow element, and expressed the loss as $CV^2/2$, where C is "related to the area ratio and whether the velocity is average or maximum".

Partington[3] expressed the kinetic energy correction in terms of viscosity, in an equation of the form:

$$\mu = \mu' - C_1\rho Q/l \qquad (13.2)$$

113

where μ' is the value of the viscosity calculated using Equation 13.1 and C_1 is a constant.

To correct for friction at the vicinity of the ends of a tube, Couette[8] added a fictitious length, Λ, to the tube length.

Stanton[9] considered that "some of the pressure difference is required to communicate the kinetic energy of motion to the water" and "that accelerations parallel to the axis of the pipe are taking place", particularly near the inlet end. He added corrections for these effects into an equation of the form:

$$\mu = (B'\Delta P/Q + C_2 \rho Q)/(l + na) \tag{13.3}$$

where a is the radius of the pipe and B', C_2, and n are constants.

We now demonstrate that all of these representations of corrections can be combined into a single corrected form of the Poiseuille equation. The Hagenbach,[4] Reynolds,[5] Wilberforce,[6] and Fain[7] representations of the correction for kinetic energy are each of the form CV^2/g in terms of pressure head loss or, more conveniently, of the form $C\rho V^2$ in terms of differential pressure.

The Poiseuille equation corrected for kinetic energy is therefore, in each case, of the form:

$$Q = B'(\Delta P - C\rho V^2)/\mu l \tag{13.4}$$

$$\mu = B'(\Delta P - C\rho V^2)/Ql \tag{13.5}$$

or

$$\mu = \mu' - B'C\rho V^2/Ql \tag{13.6}$$

In Equation 13.6, $\mu' = B'(\Delta P/Ql)$ is the value of the viscosity calculated using the uncorrected Poiseuille equation.

We now substitute $V = Q/A$, where A is the cross-sectional area of the laminar element, and we represent $B'C/A^2 l$ by the constant K. Equation 13.6 becomes

$$\mu = \mu' - K\rho Q \tag{13.7}$$

or

$$\mu = \mu' - K\dot{m} \tag{13.8}$$

where $\dot{m} = \rho Q$ is the mass flow rate through the laminar element.

Equation 13.8 becomes, after substituting $B'\Delta P/Ql$ for μ' and B for B'/l and rearranging

$$Q = B(\Delta P/\mu)/[1 + K\dot{m}(/\mu)] \tag{13.9}$$

Equation 13.2, the Partington[3] representation, readily transforms through Equation 13.8 to Equation 13.9.

Couette's transformation of l into $l + \Lambda$ is of the same form as $l + na$ in Stanton's[9] representation, Equation 13.3. Equation 13.3 can be expressed as

$$\mu = (B'\Delta P/Q + C_2\dot{m})/l(1 + na/l) \tag{13.10}$$

By making the substitutions $B'/l(1 + na/l) = B$, and $C_2/l(1 + na/l) = -K$, Equation 13.10 transforms to Equation 13.9.

DETERMINATION OF COEFFICIENTS

The volume flow rate through a laminar flowmeter can thus be expressed by an equation of the form of Equation 13.9, in which ΔP is the measured pressure drop across the laminar element and the gas viscosity, μ, and the gas density, ρ, are calculated.[10-13]

In the context of the calibration of a laminar flowmeter, an experimental determination of K in Equation 13.9 must be made for a range of values of $\dot{m}/\mu$. A range of values of $\dot{m}/\mu$ corresponds to a range of the Reynolds number.

In a calibration program, measured values of Q and ΔP and calculated values of μ and ρ provide sets of values to be inserted into a set of equations of the form of Equation 13.9. The ratio of any pair of these equations provides a first estimate of K, K. For example, the ratio of $Q_2 = B(\Delta P_2/\mu_2)/[1 + K(\dot{m}_2/\mu_2)]$ to $Q_1 = B(\Delta P_1/\mu_1)/[1 + K(\dot{m}_1/\mu_1)]$ is

$$(Q_2/Q_1) = (\Delta P_2/\mu_2)[1 + K(\dot{m}_1/\mu_1)]/(\Delta P_1/\mu_1)[1 + K(\dot{m}_2/\mu_2)] \tag{13.11}$$

When Equation 13.11 is solved for K to provide a first estimate of K, K,

$$K = [Q_2(\Delta P_1/\mu_1) - Q_1(\Delta P_2/\mu_2)]/$$

$$[Q_1\Delta P_2/\mu_2)(\dot{m}_1/\mu_1) - Q_2(\Delta P_1/\mu_1)(\dot{m}_2/\mu_2)] \tag{13.12}$$

This is the first estimate of K, most probably not the best estimate. The optimum value of K is ascertained by an iterative process in which the expression for K in Equation 13.12 and the experimental values of Q, $\Delta P/\mu$, and $\dot{m}/\mu = \rho Q/\mu$ are used in fitting a linear equation:

$$Q = D + E(\Delta P/\mu)/[1 + K(\dot{m}/\mu)] \tag{13.13}$$

From the fitting, the values of the coefficients D and E and the estimate of the standard deviation of the residuals, RSD, are recorded. A residual is the value of Q calculated using Equation 13.13 less the corresponding experimental value of Q from the calibration program.

Then, the first estimate of K from Equation 13.12 is increased by 1%, that is to $K \times (1 + 0.01n)$, where $n = 1,2,\ldots,j,j+1$, and the fitting procedure is repeated. If the RSD for n = 1 is found to be greater than that corresponding to the first estimate of K, the sign in (1 + 0.01n) is changed to negative.

The fitting procedure is repeated until the value of the RSD for n = j + 1 exceeds the value of the RSD for n = j. The optimum estimate of K (and of D and E) and the corresponding value of the RSD are those for n = j. If necessary, the optimum estimate of K from the first iteration procedure can be used as the first estimate for a second iteration procedure using smaller values of K, e.g., 0.1% K.

Although the optimum values of K, D, and E in Equation 13.13 are ascertained from the calibration data, this form of the equation cannot be used to subsequently calculate Q from measured values of ΔP and corresponding calculated values of μ and ρ since the presence of $\dot{m}$ ($\dot{m} = \rho Q$) makes an equation of the form of Equation 13.13 quadratic in Q.

Consequently, the equation to be used to calculate Q has the form of the solution of a quadratic equation:

$$Q = -1/2 \; (\mu/K\rho) \pm \{1/4 \; (\mu/K\rho)^2 + (\mu/K\rho)[A + B(\Delta P/\mu)]\}^{1/2} \quad (13.14)$$

ILLUSTRATION OF THE ITERATION PROCEDURE

The iteration procedure is illustrated by the computed results for a set of calibration data for a commercially available laminar flowmeter. The data set consists of 19 points: 5 for air, 4 for nitrogen, 5 for helium, and 5 for argon. The volume flow rate ranged between 9.5286×10^{-6} m^3 s^{-1} and 6.3561×10^{-5} m^3 s^{-1}.

The optimum value of K after the first iteration procedure, 89 1% intervals in the negative direction, was 2.570×10^{-4} m^{-1}; the corresponding values of D, E, and RSD were, respectively, -3.4692×10^{-8} m^3 s^{-1}, 1.3624×10^{-15} m^3, and 2.2×10^{-8} m^3 s^{-1}.

The optimum value of K after the second iteration procedure, 41 0.1% intervals, was 2.676×10^{-4} m^{-1}; the corresponding values of D, E, RSD, and the relative RSD (the ratio of RSD to the mid-range value of Q) were, respectively, -3.8058×10^{-8} m^3 s^{-1}, 1.3628×10^{-15} m^3, 2.1×10^{-8} m^3 s^{-1}, and 0.055%.

For comparison, the same set of calibration data was fitted to a linear equation with K = 0. The relative RSD was 0.47%, which was larger than the corresponding value above by a factor of about 8.5.

SUMMARY

An equation has been developed which relates volume flow rate (Q) through a laminar flowmeter to pressure drop (ΔP) across the flow element, the viscosity of the flowing gas (μ), the density of the flowing gas (ρ), and implicitly the dimensions of the laminar element.

It has been shown that corrections, proposed in the literature as early as 1860, to the Poiseuille equation to account for kinetic energy and end effects result in an equation of the form

$$Q = D + E(\Delta P/\mu)/[1 + K(\dot{m}/\mu)] \quad (13.15)$$

where $\dot{m} = \rho Q$ is the mass flow rate.

The best estimates of the values of D, E, and K are determined by an iterative procedure beginning with a first estimate of the value of K. A set of calibration data for a laminar flowmeter calibrated using dried air, nitrogen, helium, and argon has been used to illustrate the iterative procedure. Using Equation 13.15 to calculate Q for this set of data, the relative residual standard deviation at midrange was found to be 0.055%, which is lower by a factor of 8.5 than the corresponding quantity obtained by setting K equal to zero.

An equation to be used in calculating Q, Equation 13.14, is presented which is in the form of the solution of a quadratic equation in Q, since $\dot{m}$ in Equation 13.13 is equal to ρQ. Equation 13.14 can be used to calculate Q in the use of laminar flowmeters with flow of various gases. In addition to the gases used in the example, Equation 13.14 would presumably be applicable for other well-behaved gases over the calibration ranges of the flowmeter.

REFERENCES

1. **Poiseuille, J. L. M.,** Experimental researches on the movement of liquids in tubes of small diameter, *Compt. Rend.*, 11, 1041, 1841.
2. **Jones, F. E.,** Application of the Poiseuille equation to the treatment of laminar flowmeter calibration data, *Ind. Metrol.*, 2, 91, 1992.
3. **Partington, J. R.,** *An Advanced Treatise on Physical Chemistry*, Vol. 1, Longmans, Green, London, 1949.
4. **Hagenbach, E.,** The determination of the discharge of some fluids through the outflow from pipes, *Ann. Phys.(Leipzig)*, 109, 385, 1860.
5. **Reynolds, O.,** *Papers on Mechanical and Physical Subjects*, Vol. 2, Cambridge University Press, Cambridge, 1901.
6. **Wilberforce, L. R.,** On the calculation of the coefficient of viscosity from its rate of flow through a capillary tube, *Phil. Mag.*, 31, 407, 1891.
7. **Fain, D. E.,** Calibration of a laminar flowmeter, in *Flow, its Measurement and Control in Science and Industry*, Vol. 2, Instrument Society of America, Research Triangle Park, NC, 1981, 707.
8. **Couette, M.,** Studies on the friction of fluids, *Ann. Chim.*, 21, 433, 1890.
9. **Stanton, T. E.,** *Friction*, Longmans, Green, and Co., London, 1923, 9.
10. **Jones, F. E.,** The air density equation and the transfer of the mass unit, *J. Res. Natl. Bur. Stand.*, 83, 419, 1978.
11. **Jones, F. E.,** Calculation of compressibility factor for air over ranges of pressure, temperature, and relative humidity of interest in flowmeter calibration, U.S. Natl. Bur. Stand. NBSIR 83-2652, U.S. Government Printing Office, Washington, DC., 1983.
12. **Jones, F. E.,** Interpolation formulas for viscosities of six gases: air, nitrogen, carbon dioxide, helium, argon, and oxygen, *U.S. Natl. Bur. Stand. Technical Note 1186*, U.S. Government Printing Office, Washington, DC, 1984.
13. **Hilsenrath, J., Beckett, C. W., Benedict, W. S., Fano, L., Hoge, J. H., Masi, J. F., Nuttall, R. L., Touloukian, Y. S., and Woolley, H. W.,** Table of thermal properties of gases, *U.S. Natl. Bur. Stand. Circular 564*, U.S. Government Printing Office, Washington, DC, 1955.

Chapter 14

EFFECT OF KINEMATIC VISCOSITY ON PERFORMANCE OF TURBINE FLOWMETERS

INTRODUCTION

A turbine flowmeter consists of a turbine element, inside a housing, the angular velocity of which is essentially proportional to volume flow rate of liquid in the present discussion. The rotations of the turbine element are picked up by an electromagnetic pickup and a pulse rate or frequency is recorded.

The turbine flowmeter can be a very precise flow measurement device. However, as calibration data are conventionally treated, calibration curves are generally nonlinear. Also, the performance of the turbine flowmeter is affected by the kinematic viscosity, ν, of the flowing liquid. Jones[1] has shown that the linearity can be very significantly improved by taking into account the intercept of the frequency vs. volume flow rate relation. The linear range of the flowmeter is extended (rather, revealed) by the treatment. In this chapter, the effect of kinematic viscosity is investigated and corrected for very effectively. Figure 14.1 is a view of a typical flowmeter.

APPLICATION OF DIMENSIONAL ANALYSIS

We now discuss rather briefly dimensional analysis as it has been applied to the operation of the turbine flowmeter. The discussion begins with an influential paper by Hochreiter.[2]

Hochreiter stated that "for any device in which fluid produces a continuous motion, and where there is no significant bearing or gear friction, or shaft or other power input or output, the following dependency statement can be made, for incompressible flow Q depends on n, D, ρ, μ ...," where Q is volume flow rate (L^3T^{-1}), n is rotor angular velocity (T^{-1}) in the turbine flowmeter case, D is meter bore diameter (L), ρ is fluid density (ML^{-3}), and μ is fluid viscosity ($ML^{-1}T^{-1}$). In the parentheses the Mass, Length, and Time dimensions are given for the various quantities.

Hochreiter then applied the Buckingham Π theorem[3,4] to arrive at two dimensionless groups (five variables — three dimensions) involving the variables. Hochreiter concluded that

$$Q/nD^3 \text{ depends on } nD^2/\nu$$

and that "for incompressible flow in a turbine meter free of nonfluid friction, defining Π_1 as the flow coefficient C and Π_2 as the viscosity parameter N

$$C = Q/nD^3, \quad N = nD^2/\nu \ldots$$

FIGURE 14.1.   Phantom view of a typical turbine flowmeter. *(Photo courtesy of EG&G Instruments, Flow Technology.)*

becomes the simplest and most practical flow equation for all correlation and flow calculation purposes, and C may be correlated against N in just the same way that conventional orifice coefficients are correlated against Reynolds number." $v(\mu/\rho)$ is the kinematic viscosity of the fluid (L^2T^{-1}).

Conventionally, then,

$$\phi(Q/nD^3, nD^2/v) = 0 \tag{14.1}$$

and

$$Q/nD^3 = \phi'(nD^2/v) \tag{14.2}$$

where ϕ and ϕ' are functions.

Hochreiter assumed that the relationship between Q and n involved an *explicit* dependence on μ and ρ or on v. The general solutions of Hochreiter were then to be applied to turbine flowmeter *performance*.

In the range of Q of interest for *accurate measurements* for a particular turbine flowmeter, the relationship between a measured frequency, f, and Q are *linearly* related:[1]

$$f = a + bQ \qquad (14.3)$$

Incidentally, Hochreiter, in response to a comment, referred to the low-flow (low-speed) regions: "It is the author's position, however, that the low-speed regions, where such deviations with viscosity variations are not negligibly small, lie outside the useful range of the meter and should not be used in high accuracy work."

The intercept, a, in Equation 14.3 is not necessarily equal to zero. It is also not, of course, the frequency corresponding to $Q = 0$. The frequency of interest in the performance of the meter in the linear range is *not* f, the measured frequency; rather, it is $(f - a)$. Thus,

$$(f - a) = bQ \qquad (14.4)$$

and $(f - a)$ is *proportional* to Q.

Returning now to the first two equations in the discussion of the Hochreiter paper, we examine the group Q/nD^3 which we set equal to a dimensionless constant, κ:

$$Q/nD^3 = \kappa \qquad (14.5)$$

In the linear region of a particular meter we substitute the frequency of interest, $(f - a)$, for n in Equation 14.5:

$$Q/[(f - a)D^3] = \kappa \qquad (14.6)$$

which, when rearranged, becomes

$$(f - a) = Q/\kappa D^3 \qquad (14.7)$$

which is of the same form as Equation 14.4. D is a characteristic linear dimension for the meter; Hochreiter defined it as the meter bore diameter. Therefore, $1/\kappa D^3$ is considered to be constant in the linear region and corresponds to *b* in Equation 14.4. Consequently, the group Q/nD^3, when set equal to a dimensionless group, expresses the relationship between $n = (f - a)$ and Q in the linear range.

It has been shown experimentally that there is a dependence of turbine flowmeter performance on ν, in the linear range and in the nonlinear range. Several years after the Hochreiter paper was published, Shafer[5] published a paper in which he stated Hochreiter's general relationship:

$$Q/nD^3 = \phi(nD^2/\nu) \qquad (14.8)$$

and developed another relationship:

$$f/Q = \phi_1(f/v) \tag{14.9}$$

and said that "if this relation is correct, a plot of f/Q versus f/v would give a smooth curve the shape of which constitutes an empirical determination of the function ϕ_1 throughout the working range of the particular meter investigated." This relation and plot are in common use at the present time. The plot is commonly referred to as a "universal viscosity curve".

We now examine the left side of Equation 14.9 in light of Equation 14.3:

$$f = a + bQ \tag{14.3}$$

In the linear region of performance of the flowmeter, dividing both sides of Equation 14.3 by Q results in

$$f/Q = a/Q + b \tag{14.10}$$

that is, f/Q is *Q dependent* and a plot of f/Q against f or Q or f/v is curved because of the a/Q term — because the possibility of the existence of a nonzero intercept is ignored.

CURVATURE OF A PLOT OF f/Q AGAINST f

In this section, data from a calibration of a turbine flowmeter will be used to illustrate the determination of the linear region of the flowmeter.

Conventionally, the relationship between volume flow rate and frequency (or pulse rate) for turbine flowmeters is expressed by the "K factor".

The K factor is defined as the number of pulses per unit volume:

$$K = f/Q \tag{14.11}$$

where f is the pulse rate in hertz and Q is the volume flow rate.

A plot of K against f (or Q) is typically curved,[1] again since f/Q is Q dependent, that is,

$$K = f/Q = a/Q + b \tag{14.12}$$

Thus, the K factor is Q dependent and a plot of f/Q against f (or Q) is curved due to the presence of the a/Q term — due to the existence of a nonzero intercept.

Data for a calibration of a turbine flowmeter using type II hydrocarbon (MIL-C-7024B) illustrates some of the points made above.[1] A plot of values of f/Q against f exhibits curvature over nearly the entire range of flow rates (or f).[1]

The values of f and Q have been fitted, by linear least squares, to a linear equation beginning with the points at the six highest flow rates and continuing down to include the points at the nine highest flow rates. On the basis of the results of the fitting, the two lowest points are considered to be in the nonlinear range of the flowmeter. An algorithm to ascertain the linear region has been presented by Jones.[1]

The fitted equation for the particular turbine flowmeter is

$$f = -78.105 + 1.1139 \times 10^5 \, Q \qquad (14.13)$$

where f is in hertz and Q is in liters/second. The intercept, -78.105 hertz, was then subtracted from the values of f in the data. The mean of the nine values of $(f - a)/Q$ is 1.1140×10^5 pulses/liter, the estimate of standard deviation (SD) is 3.0×10^2 pulse/liter, and the estimate of relative standard deviation of the mean (RSDM) is 0.091%

Therefore, in the range of $180 \le f \le 1800$ Hz (a "turndown" ratio of 10), the constant value $(f + 78.105)/Q = 1.1139 \times 10^5$ pulses/liter can be used, with a corresponding RSDM of 0.091%. By contrast, the mean of the nine values of f/Q is 9.6379×10^4 pulses/liter, the SD is 1.0×10^4 pulse/liter, and the RSDM is 3.7%.

As a further illustration, water calibration data for a turbine flowmeter were treated. The 27 data points have been fitted to a linear equation. The fitted equation is

$$f = 35.123 + 1600.1 \, Q \qquad (14.14)$$

where f is in hertz and Q is in liters/second. The mean of the 27 values of $(f - 35.123)/Q$ is 1600.5 pulses/liter, the SD is 1.9 pulses/liter, and the RSDM is 0.023%. By contrast, the mean of the 27 values of f/Q is 1660.8 pulses/liter, the SD is 82 pulses/liter, and the RSDM is 0.95%. The $(f - a)/Q$ representation brings *all* 27 points, $219 \le f \le 6160$ Hz (a "turndown" ratio of 28.1), into the linear region.

Therefore, this treatment *reveals* the wide range of linearity (and the excellent precision) of the flowmeter, whereas the f/Q representation *conceals* it.

EXPERIMENTAL DETERMINATIONS OF a(v) AND b(v)

It can be argued on dimensional grounds that a could depend on v and that b could depend on (1/v). We shall not pursue that argument; rather we shall show with calibration data for three laminar flowmeters that both a and b are functions of v and that we can therefore assign the dependence of the performance of these (and other) turbine flowmeters on v to a and b, that is,

$$f = a(v) + b(v)Q \qquad (14.15)$$

Using calibration data for three turbine flowmeters, we experimentally determine a(v) and b(v), that is, the functional dependence of the parameters

TABLE 14.1
Parameters and Measures of Precision for Flowmeter A1

v (Cs)	n	a (Hz)	b (Hz/m³/s)	Mean (Q_c − Q) (m³/s × 10⁴)	SD (m³/s × 10⁴)	SD/ mid Q
0.84	30	0.076	82667.37	−0.0005	0.038	0.030%
1.18	30	1.062	82642.16	0.0000	0.057	0.045%
8.53	21	10.618	82354.39	0.0000	0.052	0.032%
18.19	15	19.278	82586.22	0.0000	0.085	0.046%

in the equation relating f to Q on the kinematic viscosity. Two of the flowmeters, designated A1 and A2, are from the same manufacturer. The other flowmeter, designated B1, is from a different manufacturer.

TREATMENT OF CALIBRATION DATA FOR FLOWMETER A1

Turbine flowmeter A1 had a diameter of 1 inch and had an RF pickoff. The pulse frequency for the calibration ranged from 103.60 to 2069.54 hertz; the flow rate ranged from 1.2085×10^{-3} to 25.0846×10^{-3} m³/s; the values of v were 0.84, 1.18, 8.53, and 18.19×10^{-6} m²/s (1 centistoke, Cs, = 10^{-6} m²/s).

For each of the values of v *in Cs*, the calibration data were fitted to an equation of the form of Equation 14.3. The resulting values of a and b are given in Table 14.1 along with the mean residuals (calculated Q − measured Q), SD, and relative SD (SD/midrange value of Q).

We note from the table that the intercept, *a*, increases monotonically with v and that the slope, *b*, generally decreases with v. We now make a linear fit of *a* to v. The resulting least-squares-fitted equation is

$$a_c = -0.159936 + 1.10208 \, v \qquad (14.16)$$

where a_c subsequently becomes the calculated value of a.

We now make a linear fit of *b* to (1/v). The resulting least-squares-fitted equation is

$$b_c = 82462.13 + 181.721 \, (1/v) \qquad (14.17)$$

where b_c subsequently becomes the calculated value of b.

Inserting a_c from Equation 14.16 and b_c from Equation 14.17 into an equation of the form of inverted Equation 14.3,

$$Q_{c4} = (f + 0.159936 - 1.10208 \, v)/[82462.13 + 181.721 \, (1/v)] \qquad (14.18)$$

where Q_{c4} is the calculated value of Q using this equation for the four values of v and subsequently for interpolated values of v.

The mean residuals, $Q_{c4} - Q$, using Equation 14.18 for the data for the four experimental values of v were −9.91, 2.64, 7.22, and 17.5×10^{-6} m²/s (from

TABLE 14.2
Parameters and Measures of Precision for Flowmeter A2

v (Cs)	n	a (Hz)	b (Hz/m³/s)	Mean (Q$_c$ – Q) (m³/s × 10⁴)	SD (m³/s × 10⁴)	SD/ mid Q
0.84	21	–0.097	83266.42	–0.0005	0.023	0.015%
1.18	21	0.006	83302.50	0.0000	0.025	0.016%
8.53	21	12.292	82789.33	0.0000	0.047	0.031%
18.19	12	31.504	82568.18	0.0000	0.066	0.033%

lowest to highest value of v). The corresponding values of SD were 4.0, 6.1, 11, and 10×10^{-6} m³/s, respectively. The corresponding values of the ratio of the SD to the midrange value of Q were 0.030, 0.048, 0.072, and 0.058%, respectively; and the corresponding values of the ratio of the mean residual to the midrange value of Q were –0.075, 0.020, –0.046, and 0.096%, respectively.

The mean residual for all 96 measurements was -1.1×10^{-6} m³/s; the corresponding value of SD was 12×10^{-6} m³/s; the corresponding value of the ratio of SD to the midrange value of Q was 0.094%; and the corresponding value of the ratio of the mean residual to the midrange value of Q was –0.0085%. These values are tabulated in Table 14.4.

TREATMENT OF CALIBRATION DATA
FOR TURBINE FLOWMETER A2

Turbine flowmeter A2 had a diameter of 1 inch and an RF pickoff. The pulse frequency for calibration ranged from 504.20 to 2095.84 hertz; the flow rate ranged from 6.05735×10^{-3} to 25.1252×10^{-3} m³/s; and the values of v were 0.84, 1.18, 8.53, and 18.19×10^{-6} m²/s.

For each of the values of v *in Cs*, the calibration data were fitted to an equation of the form of Equation 14.3. The resulting values of a and b are given in Table 14.2 along with the mean residuals, SD, and relative SD.

We note from the table that the intercept, a, increases monotonically with v and that the slope, b, generally decreases with v. We now make a linear fit of a to v. The resulting least-squares-fitted equation is

$$a_c = -2.19374 + 1.82602 \, v \tag{14.19}$$

where a_c subsequently becomes the calculated value of a.

We now make a linear fit of b to (1/v). The resulting least-squares-fitted equation is

$$b_c = 82645.85 + 607.664 \, (1/v) \tag{14.20}$$

where b_c subsequently becomes the calculated value of b.

Inserting a_c from Equation 14.19 and b_c from Equation 14.20 into an equation of the form of inverted Equation 14.3,

$$Q_{c4} = (f + 2.19374 - 1.82602 \text{ v})/[82645.85 + 607.664 (1/\text{v})] \quad (14.21)$$

where Q_{c4} is the calculated value of Q using this equation for the four values of v and subsequently for interpolated values of v.

The mean residuals using Equation 14.21 for the data for the four experimental values of v were -12.1, 26.1, 0.05, and -20.2×10^{-6} m³/s (from lowest to highest value of v); the corresponding values of SD were 8.0, 11, 7.6, and 8.0×10^{-6} m³/s, respectively; the corresponding values of the ratio of the SD to the midrange value of Q were 0.052, 0.071, 0.047, and 0.041%, respectively; and the corresponding values of the ratio of the mean residual to the midrange value of Q were -0.078, 0.17, 0.00031, and -0.10%, respectively.

The mean residual for all 75 measurements was $+0.71 \times 10^{-6}$ m³/s; the corresponding value of SD was 19.6×10^{-6} m³/s; the corresponding value of the ratio of SD to the midrange value of Q was 0.12%; and the corresponding value of the ratio of the mean residual to the midrange value of Q was 0.0046%. These values are tabulated in Table 14.4.

TREATMENT OF CALIBRATION DATA FOR FLOWMETER B1

Flowmeter B1 had a diameter of 1¼ inches and had an RF pickoff. This flowmeter was of a different manufacturer than A1 and A2. The pulse frequency ranged from 140.268 to 1521.548 hertz; the flow rate ranged from 4.0428×10^{-3} to 43.7693×10^{-3} m³/s; and the values of v were 0.92, 6.20, 10.40, 12.60, 15.40, and 20.48×10^{-6} m²/s.

For each of the values of v *in Cs*, the calibration data were fitted to an equation of the form of Equation 14.3. The resulting values of a and b are given in Table 14.3 along with the mean residuals, SD, and relative SD.

We note from the table that the intercept, a, decreases monotonically with v from v of 6.20 to 20.48×10^{-6} m²/s; however, for $\text{v} = 0.92 \times 10^{-6}$ m²/s the value of *a* is much too small to be considered to be in the sequence for the other values of v. The value of b increases monotonically with v for all of the values of v; however, for $\text{v} = 0.92 \times 10^{-6}$ m²/s the value of b is too small to be considered to be in the sequence for the other values of v. Excluding the $\text{v} = 0.92 \times 10^{-6}$ data, the values of a and b can be fitted linearly with v.

We now make a linear fit of a to v. The least-squares-fitted equation is

$$a_c = 7.07982 - 0.229132 \text{ v} \quad (14.22)$$

where a_c subsequently becomes the calculated value of a.

We now make a linear fit of b to v. The resulting least-squares-fitted equation is

$$b_c = 34273.21 + 36.6512 \text{ v} \quad (14.23)$$

where b_c subsequently becomes the calculated value of b.

TABLE 14.3
Parameters and Measures of Precision for Flowmeter B1

v (Cs)	n	a (Hz)	b (Hz/m³/s)	Mean ($Q_c - Q$) (m³/s × 10⁴)	SD (m³/s × 10⁴)	SD/ mid Q
0.92	20	2.6260	34183.94	0.0000	0.10	0.043%
6.20	20	5.5928	34486.43	0.0000	0.43	0.18%
10.40	20	4.7502	34669.03	−0.0005	0.47	0.20%
12.60	20	4.3162	34737.73	0.0000	0.44	0.19%
15.40	20	3.4405	34843.78	0.0000	0.32	0.14%
20.48	20	2.3875	35014.35	0.0000	0.30	0.13%

TABLE 14.4
Combined Results for the Several Values of v for Each of the Flowmeters

Meter No.	m (no. of v's)	Mean ($Q_{cm} - Q$) (m³/s × 10⁶)	SD (m³/s × 10⁶)	SD/mid Q	Mean ($Q_{cm} - Q$) /mid Q
A1	4(96)	−1.13	12	0.094%	−0.0085%
A2	4(75)	0.71	19	0.12%	0.0046%
B1	5(100)	−0.05	40	0.17%	−0.00025%

Note: The number of measurements, n, is in parentheses.

Inserting a_c from Equation 14.22 and b_c from Equation 14.23 into an equation of the form of inverted Equation 14.3,

$$Q_{c5} = (f - 7.07982 + 0.229132\ v)/[34273.21 + 36.6512\ v] \quad (14.24)$$

where Q_{c5} is the calculated value of Q using this equation for the five highest values of v (i.e., excluding $v = 0.92 \times 10^{-6}$ m²/s) and subsequently for interpolated values of v.

The mean residuals using Equation 14.24 for the five experimental values of v were −8.83, 8.59, 4.86, 5.19, and −4.63 × 10⁻⁶ m³/s (from lowest to highest value of v); the corresponding values of SD were 43, 47, 44, 31, and 30 × 10⁻⁶ m³/s, respectively; the corresponding values of the ratio of the SD to the midrange value of Q were 0.18, 0.20, 0.19, 0.13, and 0.13%, respectively; and the corresponding values of the ratio of the mean residual to midrange value of Q were −0.037, 0.036, 0.020, 0.022, and −0.020%, respectively.

The mean residual for all 100 measurements was −5 × 10⁻⁸ m³/s; the corresponding value of SD was 40 × 10⁻⁶ m³/s; the corresponding value of the ratio of SD to the midrange of Q was 0.17%; and the corresponding value of the ratio of the mean residual to the midrange value of Q was −0.00025%. These values are tabulated in Table 14.4.

DISCUSSION AND CONCLUSIONS

The "linearization" process described by Jones[1] delineates the region of linear performance of the turbine flowmeter and determines the parameters a and b in the linear equation. It then, and only then, becomes possible to systematically investigate the performance of the flowmeter in the linear region, the region of precise application of the flowmeter.

Calibration data for three turbine flowmeters for a number of values of kinematic viscosity, v, of liquid calibrating fluids were used to illustrate the investigation of the effects of kinematic viscosity. The performance dependence on v was assigned to a and b, i.e., a(v) and b(v).

For all three flowmeters, a was found to be linear with v. For the two flowmeters of the same manufacturer, A1 and A2, b was found to be linear with $(1/v)$. For the other flowmeter, B1, both a and b were found to be linear with v for values of v of 6.20 through 20.48×10^{-6} m²/s. At $v = 0.92 \times 10^{-6}$ m²/s, a and b lay below the fitted lines and were subsequently not included in the analysis. In this case, to investigate the v dependence in the lower region of v, it would be necessary to have data points in the 0.92 to 6.20×10^{-6} m²/s region. In the absence of such points, the analysis was restricted to the upper region for purposes of illustration.

Whether the performance of B1 was idiosyncratic of that particular flowmeter or is typical of the particular model could not, of course, be determined.

The fit of calibration values of Q to the equations is in all cases very satisfactory indeed. $(Q_c - Q)$/midrange Q ranged from 0.030 to 0.20% for the individual runs for all of the flowmeters, and ranged from 0.094 to 0.17% for all of the measurements for each flowmeter, both on an absolute basis and compared with a calibration uncertainty of 0.25%.

The "linearization" of Jones[1] has been shown for the cases studied here to be superior to the conventional K-factor approach. Also, the linearization followed by the assignment of v dependence to the parameters a and b is a very satisfactory solution to the v-dependence problem, superior to the "universal viscosity curve" approach in which the K-factor is involved.

REFERENCES

1. **Jones, F. E.,** Algorithm for ascertaining linear range of turbine flow meters, *Rev. Sci. Instrum.*, 56, 1829, 1985.
2. **Hochreiter, H. M.,** Dimensionless correlation of coefficients of turbine-type flowmeters, *Trans. ASME*, 80, 1363, 1958.
3. **Buckingham, E.,** On physically similar systems; illustrations of the use of dimensional equations, *Phys. Rev.*, IV, 345, 1914.
4. **Buckingham, E.,** Model experiments and forms of empirical equations, *Trans. ASME*, 37, 263, 1915.
5. **Shafer, M. R.,** Performance characteristics of turbine flowmeters, *Trans. ASME J. Basic. Eng.*, 83, 1, 1961.

Chapter 15

ANALYSIS OF CALIBRATION DATA
FOR VORTEX SHEDDING FLOWMETER

INTRODUCTION

In a preliminary investigation of calibration data for vortex-shedding flowmeters, the data for the calibration of a 2-inch flow transducer was analyzed.

In Table 15.1 are listed values of f, pulses/sec; Q, volume flowrate in gal/sec; $\Delta f/\Delta Q$ (ratio of the change in f to the change in Q); and f/Q.

ANALYSIS

From an inspection of Table 15.1, it is seen that the last value of $\Delta f/\Delta Q$ appears to be too low. Thus, the first eight (f,Q) pairs were fitted by the method of linear least squares to a linear equation of the form:

$$f = a + bQ \tag{15.1}$$

which became

$$f = 0.2957 + 59.0419\ Q \tag{15.2}$$

The residuals (measured f – calculated f) for the first eight data pairs ranged from 0.18 to –0.15 pulse/sec, with an estimate of standard deviation of the residuals of 0.14 pulse/sec.

The nine (f,Q) pairs were fitted to an equation of the form of Equation 15.1, resulting in

$$f = 0.3745 + 58.9928\ Q \tag{15.3}$$

The residuals for this fit ranged from 0.20 to –0.25 pulse/sec, with an estimate of standard deviation of the residuals of 0.16 pulse/sec.

As an iteration, (f – 0.4)/Q was calculated. The mean of the nine values of (f – 0.4)/Q was 58.9890 cycles/gal, with an estimate of standard deviation of 0.067 pulse/gal. Values of Q, Q_c, were calculated using the equation

$$Q_e = (f - 0.4)/58.9890 \tag{15.4}$$

The estimate of the standard deviation of the residuals for this equation was 0.0028 gal/sec, which is 0.12% of Q at midrange Q.

TABLE 15.1
Results for Vortex Shedding
Flowmeter Calibration

f (pulses/sec)	Q (gal/sec)	Δf/ΔQ (pulses/gal)	f/Q (pulses/gal)
30.216	0.50448	—	59.895
60.181	1.0128	58.949	59.419
89.615	1.5142	58.704	59.182
109.77	1.8555	59.054	59.159
135.24	2.2886	58.809	59.093
160.19	2.7103	59.165	59.104
194.89	3.2943	59.418	59.159
220.20	3.7220	59.177	59.161
250.17	4.2377	58.115	59.034

The nine (f,Q) data pairs were fitted to an equation quadratic in f, resulting in

$$Q_e = -0.00749836 + 0.0169730\ f - 7.73908 \times 10^{-8}\ f^2 \qquad (15.5)$$

The estimate of the standard deviation of the residuals was 0.0027 gal/sec, which is 0.11% of midrange Q.

Among the conclusions that can be drawn from this brief investigation are the following:

1. For the particular flowmeter, the (f/Q) vs. f representation is inadequate; the intercept for the linear equation relating f to Q cannot be neglected. For flowmeters with negligible intercepts, the (f/Q) vs. f representation would be adequate.

2. The linear fit of the eight lowest (in f) data results in an intercept of +0.2957 pulse/sec. The (f – a)/Q = (f – 0.2957)/Q vs. f representation is a much better representation of the performance of the flowmeter, especially at low frequencies (low volume flow rates). This first estimate of the intercept results in an estimate of standard deviation of the residuals of 0.0032 gal/sec, which is 0.13% of Q at midrange.

3. For a fitted equation quadratic in f, for the nine data pairs, the estimate of the standard deviation of the residuals corresponds to 0.11% of Q at midrange.

4. The (f – a)/Q "calibration factor" is a much better choice than the currently used f/Q for this particular vortex shedding flowmeter. If, for convenience, a factor of the form (f – a)/Q is desired, a quasi-optimum value of a is easily obtained. If an equation of the form:

$$Q = (f - a)/b \qquad (15.6)$$

is acceptable, the coefficients a and b are easily obtained. If an equation of the form:

$$Q = a + bf + cf^2 \qquad (15.7)$$

is acceptable, the coefficients a, b, and c can be obtained. In this particular case, the quadratic equation provides the better fit to the nine experimental data pairs.

Chapter 16

TREATMENT OF CALIBRATION DATA
FOR VENTURI METERS

The calibration data for two universal venturi meters have been fitted to linear equations (discharge coefficient, C_D, as function of Reynolds number, Re).

The data for meter No. 1 are listed in Table 16.1. The data for each of the two tap sets were fitted to equations linear in Re. For tap set #1 the resulting equation is

$$C_D = 0.97883 + 4.0 \times 10^{-11}\ Re \qquad (16.1)$$

The calculated values of C_D using Equation 16.1, C_{Dcalc}, and the residuals, C_{Dcalc} minus measured C_D, were determined. The estimate of the standard deviation of the residuals, RSD, is 0.0024. The ratio of the RSD to the mean value of C_D, RRSD, is 0.25%.

The mean of the "measured" values of C_D is 0.97894. The estimate of relative standard deviation, the ratio of the estimate of the standard deviation to the mean of the measured values of C_D, is 0.26%.

The resulting equation for tap set #2 is

$$C_D = 0.98092 - 7.46 \times 10^{-10}\ Re \qquad (16.2)$$

The RSD is 0.0017; the RRSD is 0.17%. The mean of the "measured" values of C_D is 0.97844; the estimate of the relative standard deviation is 0.20%.

The difference between the mean "measured" values of C_D for the two tap sets is 0.0005. The data for meter No. 2 are listed in Table 16.2.

For tap set #1 the fitted equation is

$$C_D = 0.97788 + 2.034 \times 10^{-9}\ Re \qquad (16.3)$$

The RSD is 0.0017; the RRSD is 0.17%. The mean of the "measured" values of C_D is 0.98291; the estimate of the relative standard deviation is 0.30%.

For tap set #2 the equation is

$$C_D = 0.97856 + 8.91 \times 10^{-10}\ Re \qquad (16.4)$$

The RSD is 0.0017; the RRSD is 0.17%. The mean of the "measured" values of C_D is 0.98076; the estimate of the relative standard deviation is 0.19%.

The difference between the mean of the "measured" values of C_D for the two taps is −0.0022.

TABLE 16.1 Calibration Data for Universal Venturi Meter No. 1			TABLE 16.2 Calibration Data for Universal Venturi Meter No. 2		
Reynolds Number, R_n	Tap Set #1 C_D	Tap Set #2 C_D	Reynolds Number, R_n	Tap Set #1 C_D	Tap Set #2 C_D
1039000	0.9822	0.9843	973000	0.9761	0.9800
1252000	0.9815	0.9823	1007000	0.9805	0.9763
1478000	0.9812	0.9799	1194000	0.9782	0.9805
1724000	0.9750	0.9765	1243000	0.9842	0.9779
1830000	0.9808	0.9789	1372000	0.9811	0.9783
2079000	0.9754	0.9777	1419000	0.9815	0.9792
2256000	0.9804	0.9793	1647000	0.9802	0.9818
2536000	0.9745	0.9780	1726000	0.9822	0.9793
2735000	0.9786	0.9782	2024000	0.9830	0.9810
2909000	0.9764	0.9782	2118000	0.9811	0.9828
3272000	0.9803	0.9777	2281000	0.9833	0.9815
3551000	0.9789	0.9760	2504000	0.9824	0.9826
4002000	0.9768	0.9790	2759000	0.9837	0.9804
4834000	0.9784	0.9782	3112000	0.9838	0.9813
4889000	0.9806	0.9776	3179000	0.9860	0.9821
4907000	0.9770	0.9792	3620000	0.9840	0.9845
5204000	0.9794	0.9766	3822000	0.9868	0.9823
5217000	0.9809	0.9767	4061000	0.9873	0.9816
5235000	0.9815	0.9772	4468000	0.9860	0.9806
5333000	0.9794	0.9774	4869000	0.9867	0.9812

It is apparent from this treatment that the C_D's of these two devices did not depend strongly on Re in the ranges of Re for the calibration tests.

For tap set #1 of meter No. 1, the C_D increased by at most 0.015% over the range of Re; in fact, the use of the mean "measured" C_D would not significantly increase the imprecision. For tap set #2 the C_D decreased by 0.33%; it would not be appropriate to use the mean "measured" C_D for this tap set. The mean of the "measured" C_D's for the two tap sets differed by 0.05%.

For tap set #1 of meter No. 2, the C_D increased by 0.81% over the range of Re. For tap set #2 the C_D increased by 0.35%. The mean of the "measured" C_D's for the two tap sets differed by 0.22%.

The fit of the regression lines to the "measured" points, as indicated by the RRSD is 0.25% for tap set #1 of meter No. 1 and 0.17% for each of the other three, recalling that each of the points represents the mean of five (usually) "observations" of C_D and of Re.

Chapter 17

ASCERTAINING LINEAR RANGE OF ANEMOMETERS

INTRODUCTION

Baynton[1] has stated that "although true helical anemometers obey the equation S=bR, all other types of rotational anemometers obey the equation S=a+bR." S is the wind speed and R is rotational rate; a and b are constants. A least-squares analysis of wind tunnel test data to estimate a and b and a statistical test of whether the intercept a is significantly greater than zero were proposed to determine whether S=a+bR or S=bR describes the performance of a rotational anemometer. The parameters a and b would be determined from wind tunnel measurements throughout a specified working range.

It has been customary to present anemometer calibration data either in the form of a table of corrections to anemometer indicated airspeed, I; as a plot of the ratio of the wind tunnel airspeed, S, to I, against S; or a plot of (S–I)/S, against S, expressed as percent. In cases in which a, the intercept, is significantly different from 0, taking the ratio S/I generates curvature in the relationship between S/I and I, that is, much of the curvature is an artifact of taking the ratio of S to I.

JONES'S TREATMENT OF CALIBRATION DATA FOR TURBINE FLOWMETERS

Jones[2] has treated calibration data for turbine flowmeters, devices that are similar in structure and function to anemometers with rotating mechanisms, and has demonstrated the effects of taking the ratio of the dependent variable to the independent variable. In that case also, much of the curvature in the plot of the ratio against either the dependent or the independent variable is an artifact of taking the ratio.

Correcting for the nonzero intercept resulted in an extended range of linearity of the calibration function, and thus in an apparent significant improvement in the linearity of the device.

In this chapter, the method described in Reference 2 is applied to anemometer calibration data.

TREATMENT OF ANEMOMETER CALIBRATION DATA

For that portion of the airspeed, S, range for which the indication of the anemometer, I, is *linear*,

$$I = a + bS \qquad (17.1)$$

The intercept or zero offset, a, is not necessarily equal to 0; it is also not, of course, the indication corresponding to S = 0. The quantity of interest in the linear range is, consequently, (I − a).

On dividing both sides of Equation 17.1 by S,

$$I/S = a/S + b \qquad (17.2)$$

Thus, a plot of I/S against S is curved due to the presence of the a/S term (it depends on S), i.e., due to the presence of a nonzero intercept. A similar argument holds if Equation 17.1 is rearranged with S as the dependent variable.

For the linear expression, Equation 17.2, the ratio (I − a)/S is constant. For measurements, however, values of this ratio are affected by the variability of the measurements. Estimates of the values of a and b are made by linear least-squares fitting of the experimental values of I and S. It is then necessary to estimate the limits of the linear region of the working range of the anemometer.

ALGORITHMS

There are several algorithms that can be used to estimate the limits of the linear range of the anemometer. In one of these, the (I,S) data pairs are arranged in descending order of I or S, and values of $\Delta I/\Delta S$ are calculated using successive values of I and S, where ΔI and ΔS are differences. A running mean ratio of ΔI to ΔS and an estimate of standard deviation (SD) corresponding to the running mean are calculated.

The difference between the value of $\Delta I/\Delta S$ corresponding to a data point and the running mean for the preceding data points is divided by the SD corresponding to the running mean. This ratio indicates how far, in multiples of SD, the value of $\Delta I/\Delta S$ is from the running mean. When the multiple exceeds a criterion value (e.g., 3), the particular point is tentatively considered to be beyond the linear range. The value of I (or S) for the immediately preceding data point is the first estimate of the lower limit of the linear range.

Then, the points at this limit and above are fitted to a linear equation of the form of Equation 17.1 and the estimate of the residual (calculated S − measured S) standard deviation, RSD, is calculated. The RSD is a measure of fit of the experimental points to the fitted line.

Next, points are added one at a time (at the next lower I or S) to the linear analysis until the RSD increases significantly; the criterion for judging significant increase is determined for individual cases based on desired or acceptable precision or range of application. For the maximum number of points for which the RSD does not increase significantly, the values of a and b in Equation 17.1 and the RSD are the desired parameters.

APPLICATION OF THE ANALYSIS

The linear analysis has been applied to experimental data for a variety of anemometer types. The results are illustrated in Figure 17.1.

The plots in Figure 17.1 were derived from calibration data for a Biram-type anemometer with aluminum blades and miniature ball bearings on the main spindle shaft. The wind tunnel airspeed ranged from 353.7 to 4585.7 ft/min (1.797 to 23.295 m/s).

Of the 14 data points (the point at the lowest airspeed was not included), 13 were fitted by linear least squares. The resulting equation is

$$I = -10.730 + 1.06867 \, S \qquad (17.3)$$

where I is indicated airspeed and S is wind tunnel airspeed. For convenience, the original airspeed units, ft/min, are used.

In Figure 17.1, the quantity $(I + 10.730)/S$ is plotted against S for all 14 points (closed circles). Also, the quantity I/S is plotted against S (open circles). The solid horizontal line represents the slope, 1.06867, in Equation 17.3.

The residual standard deviation for the 13 estimates of S calculated using Equation 17.3 is 2.8 ft/min (0.014 m/s), which is 0.11% of the midrange value of S.

The curvature in the plot of I/S against S is readily apparent, as is the effectiveness of adjusting for the intercept in "linearizing" the output of the anemometer.

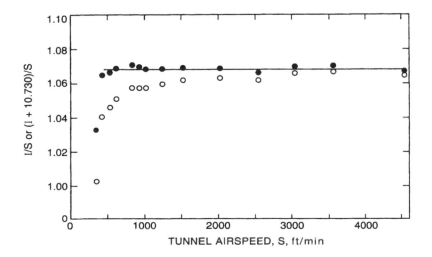

FIGURE 17.1. Plots of I/S against S (open circles) and $(I + 10.730)/S$ against S (closed circles) for calibration data for a Biram-type anemometer.

REFERENCES

1. **Baynton, H. W.,** Errors in wind run estimates from rotational anemometers, *Bull. Am. Meteor. Soc.*, 59, 1127, 1976.
2. **Jones, F. E.,** Algorithm for ascertaining linear range of turbine flow meters, *Rev. Sci. Instrum.*, 56, 1829, 1985.

Chapter 18

DIVERTER CORRECTIONS

INTRODUCTION

Flow metering devices for liquids are calibrated gravimetrically. The flow rates for the calibration are determined by collecting a measured mass of water in a tank during a measured interval of time. A timer is triggered by the actuation of a diverter valve which diverts the flow of water to the collection tank or to a reservoir. The measured time interval is generally not equal to the effective collection time. The major part of the discrepancy between the effective collection time and the measured time interval is ascribed to the diverter valve. Various methods[1,2] have been devised for calculating the diverter error or adjusting the diverter valve.

In this chapter, a method[3] developed for estimating the corrections to be made to the measured time interval for diverters will be described, and data for two diverters will be used to illustrate the calculations.

DIVERTER VALVE CORRECTION

The diverter value correction, τ, is estimated using the model embodied in the expression

$$\dot{m} = m/(t + \tau) \tag{18.1}$$

where $\dot{m}$ is the mass flow rate, m is the mass of water collected during a diversion, t is the measured collection time, and τ is the diverter valve correction. $(t + \tau)$ is the effective collection time.

To acquire data to be used in estimating the diverter valve correction, m and t are measured for each of several mass flow rates, $\dot{m}$. Also, for each nominal value of $\dot{m}$, m_i and t_i are measured for shorter time intervals ($t_i \simeq t/10$). The corresponding values of mass flow rate and diverter valve correction are $\dot{m}_i$ and τ_i.

It is assumed that the correction for the shorter time interval is the same as the correction for the longer time interval. For the shorter time intervals, an equation similar to Equation 18.1 applies:

$$\dot{m}_i = m_i/(t_i + \tau_i) \tag{18.2}$$

An auxiliary flow measurement device in the flow system is used in making estimates of $\dot{m}$ and $\dot{m}_i$. In the use of the auxiliary device for this purpose, the approximate relationship between the parameter indicated by the device (frequency or total counts in the case of a turbine flow meter) and the mass flow rate is required. A turbine flow meter, due to its excellent short-term repeatability, is a suitable choice as the auxiliary flow measurement device.

The steps followed in estimating the diverter correction are summarized in the following:

1. Using an auxiliary flow meter (a turbine flow meter in this example), set the desired nominal flow rate and monitor it.
2. Collect a mass of water, m_0, (at the nominal flow rate) during a measured time interval, t_0, while recording the frequency of the auxiliary flow meter to determine the mean frequency, f_0.
3. Repeat step 2, at least once.
4. Collect a mass of water, m_i, at the same nominal flow rate during a measured time interval, t_i, $(t_i = t_0/10)$, while recording the frequency of the auxiliary flow meter to determine the mean frequency, $(f_i)_{mean}$.
5. Repeat step 4 as many times as practicable.
6. Repeat step 2 at least twice.
7. Calculate the first estimate of C for the set of m_0, t_0, and f_0 data using Equation 18.4 below. Calculate the mean first estimate of C.
8. Insert the mean first estimate of C into Equation 18.6 below and calculate τ_i for the set of m_i, t_i, and f_i data. Calculate $(\tau_i)_{mean}$, the mean of the τ_i.
9. Insert $(\tau_i)_{mean}$ into Equation 18.7 below and recalculate C.
10. Insert the recalculated value of C into Equation 18.6 below and recalculate τ_i and $(\tau_i)_{mean}$.
11. Repeat steps 9 and 10 until C becomes essentially constant.
12. Insert the constant (asymptotic) value of C into Equation 18.6 below to make final calculation of τ_i and $(\tau_i)_{mean}$.
13. $(\tau_i)_{mean}$ is the best estimate of the value of the diverter correction at the nominal flow rate.
14. Repeat steps 1 through 12 for other nominal flow rates throughout the desired range of flow rate.

Using Equation 18.1, first estimates are made of $\dot{m}_0$ for the longer collection times, t_0, required to fill the collection tank, by setting $\tau = 0$. These estimates of $\dot{m}_0$ are used to determine the approximate relationship between $\dot{m}_0$ and the parameter indicated by the auxiliary flow measurement device:

$$\dot{m}_0 \simeq m_0/t_0 = Cf_0 \qquad (18.3)$$

$$C = m_0/t_0f_0 \qquad (18.4)$$

where f_0 is the frequency indicated by the auxiliary flow meter, for example.

C from Equation 18.4 is then used to estimate $\dot{m}_i$ from the corresponding f_i, using the relationship

$$\dot{m}_i \simeq Cf_i \qquad (18.5)$$

By substituting Cf_i for $\dot{m}_i$, and rearranging Equation 18.2, we arrive at the following expression for the diverter correction, τ_i:

$$\tau_i = (m_i/Cf_i) - t_i \qquad (18.6)$$

From measurements of m_i, f_i, and t_i made under the same nominal mass flow conditions, Equation 18.6 is used to estimate τ_i.

An iterative procedure is used to make the best estimate of τ. $(\tau_i)_{mean}$, the mean of the n values of τ_i calculated using Equation 18.6, is inserted into the following equation to recalculate C:

$$C = m_0/f_0[t_0 + (\tau_i)_{mean}] \qquad (18.7)$$

This iterative procedure is repeated until the values approach a constant (asymptotic) value. The best estimates of τ_i and $(\tau_i)_{mean}$ are determined by inserting the asymptotic value of C in Equation 18.6.

EXPERIMENTAL DETERMINATIONS OF τ

Experimental determinations of diverter valve corrections using the procedures in this chapter will now be illustrated. To determine the diverter valve corrections for two diverters, two sets of measurements were made. For one run for each of the diverters, data and the calculated diverter valve corrections are given in Tables 18.1 and 18.2.

A 946-liter-capacity collection tank with an attached diverter was used for the first set of measurements. For six mass flow rates, measurements were made of m_0, t_0, f_0, m_i, t_i, and f_i for water. The water flow through a 102-mm (4-in) turbine meter in a 102-mm (4-in) meter run was diverted into the collection tank. The data and results are summarized in Table 18.3.

The mean of the six asymptotic values of C was found to be 0.053016 kg/s/Hz with an estimate of standard error (s.e.) of 0.000016 kg/s/Hz. (The estimate of s.e. is the estimate of standard deviation divided by the square root of the number of measurements, n). For the six mass flow rates, the values of τ_i were not statistically different; therefore, the 50 values could be pooled.

The mean and the estimate of standard error of the pooled values are 0.006 and 0.003 s, respectively. The best estimate of τ over the range of $\dot{m}$ of 3.981 to 26.32 kg/s is thus 0.006 s. This value of τ is then inserted into Equation 18.1 to calculate $\dot{m}$ from measurements of m and t in the calibration of water flow rate measuring devices.

A 1820-liter-capacity collection tank with an attached diverter was used for the second set of measurements. The same turbine meter and meter run used for the first set of measurements was used for the second set. The total counts, N, from the turbine meter rather than frequency was used as the parameter. The approximate value of C then is m_0/N_0, and the expression for τ_i is

$$\tau_i = t_i[(m_i/CN_i) - 1] \qquad (18.8)$$

TABLE 18.1
Data and Calculated Diverter Valve Corrections
for One Run in the 946-Liter Tank

m_i (kg)	t_i (s)	f_i (Hz)	$\tau_i{}^a$ (s)
80.24	10.113	149.60	0.005
81.35	10.274	149.25	0.008
81.74	10.323	149.10	0.019
80.47	10.164	149.70	−0.024
79.24	9.684	149.70	0.001
81.94	10.302	149.65	0.027
80.63	10.204	149.65	−0.041
81.87	10.266	149.70	0.051
80.08	10.128	149.20	−0.006
80.15	10.119	149.20	0.015

Note: $(\tau_i)_{mean}$ = 0.0055 s. Estimate of standard error = 0.0082 s.
$^b m_0$ = 791.65 kg. $^b t_0$ = 99.926 s. $\dot{m}_0$ = 7.9224 kg/s.
$^b f_0$ = 149.44 Hz. C = 0.053011 kg/s/Hz.

^{a} Calculated using Equation 18.6.
^{b} Mean value.

(From Jones, F. E., Int. J. Heat Fluid Flow, 5, 247, 1984, with permission.)

For three mass flow rates, measurements were made of m_0, t_0, N_0, m_i, t_i, and N_i. The data and results are summarized in Table 18.4.

The mean of the three asymptotic values of C was 0.052906 kg/count with an estimate of standard error of 0.000034 kg/count. At the 12.6348 kg/s flow rate, the value of τ_i is significantly different from the other two values. Typical of diverter valves of the design of this diverter valve, the physical performance of the diverter valve is quite different for relatively low flow rates than for the higher flow rates. To fully characterize either of the two diverter valves, additional data at other flow rates of interest would be required.

CONCLUSIONS

Unwarranted assumptions are avoided by the method presented in this chapter. The major assumption, other than the obvious relationship embodied in Equation 18.1, is that the coefficient C is constant during the time (of the order of minutes) during which a set of measurements is made at a nominal flow rate. This assumption is much more realistic than the assumption of the equality $\dot{m}_0 = \dot{m}_i$, for example.

In the method, data provided by the individual points give statistical information on the variability of the performance of the system as indicated by the

TABLE 18.2
Data and Calculated Diverter Valve Corrections
for One Run in the 1820-Liter Tank

m_i (kg)	t_i (s)	N_i	$\tau_i{}^a$ (s)
184.63	14.5686	3480	0.0233
184.72	14.6154	3480	0.0305
179.46	14.2253	3389	−0.0043
179.15	14.1460	3375	0.0298
174.87	13.8464	3298	0.0139
178.09	14.1051	3360	0.0088
179.60	14.1904	3387	0.0151
179.55	14.2067	3386	0.0154
178.20	14.0826	3362	0.0091
178.70	14.1780	3377	−0.0142

Note: $(\tau_i)_{mean} = 0.0127$ s. Estimate of standard error $= 0.0044$ s.
$^b m_0 = 1788.95$ kg. $^b t_0 = 141.5892$ s. $\dot{m}_0 = 12.6348$ kg/s.
$^b N_0 = 33770.0$. $C = 0.052970$ kg/count.

^a Calculated using Equation 18.8.
^b Mean value.

(From Jones, F. E., Int. J. Heat Fluid Flow, 5, 247, 1984, with permission.)

TABLE 18.3
Summary of Data and Results
for the Measurements for the 946-Liter Tank

$(\dot{m}_0)_{mean}$ (kg/s)	$(f_0)_{mean}$ (Hz)	$(\tau_i)_{mean}$ (s)	s.e. (s)	n
3.981	75.01	0.009	0.008	10
7.922	149.44	0.005	0.008	10
11.97	225.90	0.013	0.005	9
15.97	301.06	0.004	0.005	10
20.00	377.10	−0.001	0.008	6
26.32	496.95	0.003	0.013	5

(From Jones, F. F., Int. J. Heat Fluid Flow, 5, 247, 1984, with permission.)

variability of the calculated value of τ. The estimate of standard error is used with the mean values as a measure of the precision of the mean.

Because the examples in this chapter are included to illustrate the procedures, the subject of retaining the mean diverter valve correction based on its relative magnitude compared to the longer collection time was not addressed.

TABLE 18.4
Summary of Data and Results
for the Measurements for the 1820-Liter Tank

$(\dot{m}_0)_{mean}$ (kg/s)	$(N_0)_{mean}$ (counts)	$(\tau_i)_{mean}$ (s)	s.e. (s)	n
12.6348	33770.0	0.0128	0.0044	10
25.1950	34012.2	0.0257	0.0017	9
31.4662	34154.0	0.0270	0.0006	10

(From Jones, F. E., Int. J. Heat Fluid Flow, 5, 247, 1984, with permission.)

It would, however, be good practice to retain the mean values given in Tables 18.1 and 18.2 because in each case they amount to approximately 0.01% of the longer collection time.

REFERENCES

1. **Kinghorn, F. C., MacKay, A., and Harrison, P.,** The new United Kingdom primary standards for water flowrate measurement, in *Flow: Its Measurement and Control in Science and Industry*, Vol. 2, Instrument Society of America, Research Triangle Park, NC, 1981, 761.
2. **Hayward, A. T. J.,** *Flowmeters, A Basic Guide and Source-Book for Users*, John Wiley & Sons, New York, 1979, 144.
3. **Jones, F. E.,** Estimating diverter valve corrections, *Int. J. Heat Fluid Flow*, 5, 247, 1984.

Chapter 19

CALIBRATION OF PLATFORM SCALE

INTRODUCTION

It is the purpose of this chapter to illustrate the calibration of a platform scale used to determine the mass of water in a tank used for flowmeter calibration.

Four calibration runs were made on the scale by two different operators, A and B. Calibration consisted of placing steel standard weights on the platform of the scale and reading the dial of the scale. The scale was scaled in such a way that the indications of interest fell within the approximate 109 to 129 region of the dial. Both operators read the dial indications visually at each point. There was a systematic difference between the observations of the two operators. The observations of operator A have been chosen to be treated in detail and to compute the systematic difference between the operators.

The relation between the mass of the standards, m_S, and the mass of the internal weights of the scale, m_H, can be expressed ideally as

$$m_S [1 - (\rho_a/\rho_S)] = K \, m_H [1 - (\rho_a/\rho_H)] \qquad (19.1)$$

where ρ_a is air density, ρ_S is the density (7.8 g/cm^3) of the standard weights, ρ_H is the density of the cast iron interval weight and of the hanger weights (7.15 g/cm^3) of the scale, and K is an effective lever ratio.

The practical relation between m_S and the scale indication D (D = dial reading + the sum of 200-lb nominal increments of hanger weights) can be expressed as

$$m_S [1 - \rho_a/\rho_S)] = f(D) [1 - (\rho_a/\rho_H)] \qquad (19.2)$$

where f(D) is a polynomial function of D. Then,

$$m_S = f(D) [1 - (\rho_a/\rho_H)/(1 - (\rho_a/\rho_S)] \qquad (19.3)$$

For a nominal air density of 1.17×10^{-3} g/cm^3, the ratio of $[1 - (\rho_a/\rho_S)]$ to $[1 - (\rho_a/\rho_H)]$ is 1.000014.

EXPERIMENTAL

The values of m_S and D for the four runs are listed in Table 19.1. The values for all four runs, a total of 102 number pairs, have been fitted to an equation linear in D:

$$m_{Sc} \text{ (lb)} = a + b \, D[1 - (\rho_a/\rho_H)]/[1 - (\rho_a/\rho_S)] \qquad (19.4)$$

TABLE 19.1
Scale Calibration Data

$$D(1 - \rho_a/\rho_H)/(1 - \rho_a/\rho_S), \text{ lb}$$

m_s, lb	Run 1	Run 2	Run 3	Run 4
49.999	109.60(0.10)	109.50(0)		
59.999	119.60(0.10)	119.45(−0.05)		
69.999	129.55(0.05)	129.50(0)		
250.002	309.45(0)	309.50(0.05)	309.60(0.15)	309.60(0.15)
260.002	319.45(0)	319.45(0)	319.60(0.15)	319.55(0.10)
270.002	329.45(0)	329.45(0)	329.55(0.10)	329.55(0.10)
450.004	509.34(−0.06)	509.34(−0.06)	509.54(0.14)	509.39(−0.01)
460.004	519.34(−0.06)	519.29(−0.11)	519.49(0.09)	519.24(−0.16)
470.004	529.25(−0.11)	529.24(−0.16)	529.44(0.04)	529.29(−0.11)
649.998	709.29(−0.05)	709.24(−0.11)	709.54(0.21)	709.19(−0.16)
659.998	719.24(−0.10)	719.19(−0.15)	719.49(0.15)	719.14(−0.20)
669.998	724.19(−0.15)	729.14(−0.20)	729.44(0.10)	729.14(−0.20)
850.000	909.39(0.10)	909.59(0.30)	909.24(−0.05)	909.14(−0.15)
860.000	919.39(0.10)	919.44(0.15)	919.21(−0.05)	919.04(−0.25)
870.000	929.34(0.05)	929.44(0.15)	929.19(−0.10)	929.04(−0.25)
1049.988	1109.38(0.15)	1109.43(0.20)	1109.38(0.15)	1108.98(−0.25)
1059.988	1119.23(0)	1119.38(0.15)	1119.33(0.10)	1118.88(−0.35)
1069.988	1129.28(0.06)	1129.33(0.10)	1129.23(0.01)	1128.98(−0.24)
1249.990	1309.33(0.24)	1309.18(0)	1309.48(0.39)	1308.98(−0.20)
1259.990	1319.23(0.06)	1319.13(−0.04)	1319.38(0.21)	1318.93(−0.24)
1269.991	1329.23(0.06)	1329.13(−0.04)	1329.33(0.16)	1328.93(−0.24)
1449.992	1509.33(−0.09)	1509.33(0.21)	1509.08(−0.04)	1509.13(0.01)
1459.992	1519.03(−0.09)	1519.28(0.16)	1519.08(−0.04)	1519.03(−0.09)
1469.992	1528.98(−0.14)	1529.23(0.11)	1529.13(0.01)	1529.03(−0.09)
1649.987	1708.98(−0.09)	1709.33(0.26)	1709.18(0.11)	1709.13(0.06)
1659.988	1718.88(−0.19)	1719.28(0.21)	1719.13(0.06)	1719.08(0.01)
1659.988	1728.80(−0.17)	1729.28(0.21)	1729.03(−0.04)	1729.03(−0.04)

The resulting equation is

$$m_{Sc} \text{ (lb)} = -59.533 + 1.000265 \, D[1 - \rho_a/\rho_H]/[1 - (\rho_a/\rho_S)] \qquad (19.5)$$

The residuals [calculated m_S (m_{Sc}) − known m_S] are listed in parentheses in Table 19.1. The estimate of the standard deviation of the residuals is 0.14 lb. Equation 19.5 enables calculation of the mass on the scale from observations of D.

For the case of tank calibration, the relation between the mass of the empty tank, m_T, the mass of water in the tank, m_W, and the calculated mass on the scale, m_{Sc}, is

$$m_T[1 - (\rho_a/\rho_T)] + m_W[1 - (\rho_a/\rho_W)] = m_{Sc}[1 - (\rho_a/\rho_{H,S})] \qquad (19.6)$$

where ρ_T is the density of the material of which the tank is constructed, ρ_W is the density of the water in the tank, and $\rho_{H,S}$ is the density of the mass standards and of the internal weights and hanger weights.

For the case of no water in the tank,

$$m_T = m_{Sce}[1 - (\rho_a/\rho_{H,S})]/[1 - (\rho_a/\rho_T)] \qquad (19.7)$$

where m_{Sce} is the calculated (from the value of D) mass of standards corresponding to an empty tank.

For the case of water in the tank,

$$m_W = (m_{Scf} - m_{Sce})[1 - (\rho_a/\rho_{H,S})]/[1 - (\rho_a/\rho_W)] \qquad (19.8)$$

where m_{Scf} is the calculated mass of standards with water in the tank.

The mean difference between the observations of operators A and B for 79 pairs of observations was found to be 0.03 lb.

Chapter 20

VOLUMETRIC TEST MEASURES

INTRODUCTION

Procedures for the calibration of volumetric test measures, devices used to contain or deliver known volumes of water, were described by Houser[1] with the considerable assistance of R. M. Schoonover.

The three calibration constants that must be determined in the calibration of volumetric test measures are

1. *Containment volume*, V_C, the volume of water required to fill the test measure at 60°F.
2. *Delivery volume*, V_D, the volume of water that may be poured from the test measure at 60°F under specified conditions.
3. *Neck constant*, K, relating the true volume of the neck of the test measure to the value observed on the scale of the neck.

V_C and V_D are significantly different.

TYPES OF TEST MEASURES

There are several basic types of test measure. Almost all test measures, regardless of design, have some characteristics in common. The majority of test measures have nominal volumes ranging from 1 to 1000 gallons, in increments of 5 gallons. The 1-, 5-, 50-, and 100-gallon test measures are most frequently encountered.

Nearly all test measures are fabricated from mild steel or 304 stainless steel. The coefficient of thermal cubical expansion of mild steel is $18.6 \times 10^{-6}/°F$; the coefficient for 304 stainless steel is $26.5 \times 10^{-6}/°F$. By agreement, various organizations report the volumes of test measures at 60°F.

CALIBRATION OF TEST MEASURES

The two practical methods used in the calibration of test measures are[2]

1. The *gravimetric* method.
2. The *volumetric* method.

The gravimetric method establishes the mass of water either contained or delivered by or from a test measure; the associated volume of water is inferred from the mass, using the density of water. The volumetric method employs a standard vessel of known volume from which water is transferred to the test measure being calibrated; the contained or delivered volume is derived from the water transferred from the standard vessel.

The gravimetric method requires a balance for making mass measurements. The volumetric method is essentially a counting procedure in which the volume of the test measure is determined from the number of "dumps" of the standard vessel required to fill the test measure.

Volume of test measures is defined at at least one reference point (end point); in many vessels there are two reference points.

NECK CALIBRATION

In use, the test measure is not always filled to the same scale reading, δ_i, used in the calibration of the test measure but might be filled to any scale reading, δ_n. The volume difference between the scale readings, δ_i and δ_n, is determined from neck scale calibration. The neck calibration is accomplished by filling the test measure to near the bottom of the range of the neck scale, δ_A, and inserting precision spheres of known diameter, d, to increase the water level successively to δ_B, δ_C, etc.

Effectively, water volume is added in known increments and the corresponding scale readings are recorded. The neck constant, K, is determined from analytical evaluation of the data using equations of the form:

$$K = (\pi d_B^3)/(\delta_A - \delta_B)^6$$

where d_B is the diameter of the precision sphere corresponding to δ_B.

The test measure must, of course, be level to insure a true neck calibration. For test measures with neck diameters greater than 4 inches, precision spheres are replaced by known volumes of water.

GRAVIMETRIC CALIBRATION PROCEDURE

In the gravimetric calibration procedure, the mass of water required to fill the test measure minus the mass of water clinging to the walls of the test measure after it has been drained in a specified manner is determined. The density of the water is used to calculate the containment volume, V_C, and the delivery volume, V_D. Any weighing procedure of appropriate precision can be used.

Prior to measurements, a cap is placed over the neck opening of the test measure, and remains during the entire weighing procedure, to restrict the evaporation of water from the vessel.

The empty test measure is first weighed; the air temperature, barometric pressure, and the relative humidity are recorded before and after each weighing.

The test measure is then filled with previously distilled water (equilibrated with the ambient conditions) to a point on the neck scale, δ_i. After the filling, the water is stirred thoroughly to remove thermal gradients and to allow air bubbles to rise to the surface.

The temperature, t_w, is then recorded; the test measure is leveled; the water in the gauge tube is agitated; and the location of the meniscus in the neck is read. The meniscus scale reading, the water temperature, and the mix time are then recorded. After its exterior is thoroughly dried, the filled test measure is weighed. The ambient temperature, pressure, and relative humdity and the mass are then recorded.

The water is then drained from the test measure; it is held in the drainage position for a specified interval, usually 10 to 30 seconds, after the apparent cessation of the main flow. Then the test measure is again weighed and its mass, the ambient temperature, pressure, and relative humidity and the drain time are recorded.

Gravimetric Calibration Calculation

From the weighings described above and appropriate buoyancy corrections (see Chapter 8), three masses are determined:

1. M_j, the mass of the empty test measure.
2. M_w, the mass of water contained by the test measure when filled.
3. M_{Rw}, the mass of water retained by the test measure after being drained.

The containment volume, V_C, is calculated from M_w using the water density, corresponding to the water temperature, calculated using the equations in Chapter 3.

The mass of the retained water, M_{Rw}, is then inferred from the difference in mass between the drained test measure and the empty test measure. The mass of the delivered water is the difference between M_w and M_{Rw}. The delivery volume, V_D, is calculated from the appropriate water density.

By convention, test measure volumes are referenced to 60°F. To convert from the containment volume at the calibration temperature (t), V_{Ct}, to the volume at 60°F, V_{C60}, the following equation is used:

$$V_{C60} = V_{Ct} \left[1 + \alpha(60 - t) \right] \tag{20.1}$$

where α is the coefficient of cubical thermal expansion of the metal of which the test measure is fabricated.

VOLUMETRIC CALIBRATION PROCEDURE

The second calibration procedure is referred to as the "volumetric transfer" method. In this method, a known volume of water is transferred from a standard vessel to the test measure being calibrated and either a contained or delivered volume of the test measure is calculated.

The standard vessel is leveled and filled to a predetermined point on the neck scale. The water temperature in the standard vessel, t_a, is measured and recorded. The water in the gauge tube is agitated to get a uniform meniscus and the meniscus scale reading is taken and recorded.

The test measure is leveled and placed at a lower level than the standard vessel. The standard vessel is drained into the test measure for a specified interval (usually 30 seconds from the apparent cessation of the main flow). The water in the gauge tube is agitated and the neck reading and the water temperature, t_b, are observed and recorded. When the standard vessel has a smaller volume than that of the test measure, the steps above are repeated the appropriate number of times.

After the completion of one complete calibration or "run", a second, check, run can be made. The test measure is drained for the appropriate interval before the second run is started.

If the test measure is to be calibrated for capacity, the internal surfaces of the test measure must be dry; for delivery, the internal surfaces of both the standard and the test measure must be wet. The internal surfaces are wetted by filling the standard and transferring the water into the test measure with a 30-second drain time after the apparent cessation of the main flow; the test measure is drained in a similar manner.

A neck calibration for the test measure is performed as described above.

The appropriate corrections are made for the departure of temperatures from 60°F.

EQUIVALENCE OF GRAVIMETRIC AND VOLUMETRIC TEST MEASURE CALIBRATION

Schoonover[2] has shown that the gravimetric and volumetric methods of test measure calibration may be considered to be equivalent.

EXAMPLE OF UNCERTAINTY FOR GRAVIMETRIC CALIBRATION

Schoonover[2] provided data for gravimetric and volumetric calibrations of test measures. For example, a 30-gallon test measure was calibrated gravimetrically. The mean of five determinations of the volume of the test measure was 30.00150 gallons. The type A standard uncertainty (see Chapter 7) was 0.00089 gallons; the type B standard uncertainty was 0.000447 gallons; the combined standard uncertainty was 0.0010 gallons. The expanded standard uncertainty with a coverage factor of k = 2 was 0.0020 gallons. The relative expanded uncertainty was 0.0067%.

REFERENCES

1. **Houser, J. F.,** Procedures for the Calibration of Volumetric Test Measures, National Bureau of Standards NBSIR 73-287, 1973.
2. **Schoonover, R. M.,** The Equivalence of Gravimetric and Volumetric Test Measure Calibration, National Bureau of Standards NBSIR 74-454, 1974.

Chapter 21

EXPERIMENTAL DETERMINATION OF DENSITY OF LIQUIDS

LIQUID DENSITY DETERMINATION BY HYDROSTATIC WEIGHING

Bowman and Schoonover[1] used hydrostatic weighing to determine the density of solids. In this chapter, hydrostatic weighing is applied to the determination of the density of liquids.

In hydrostatic weighing, a solid object is weighed first in air then in a liquid (usually water). Weighing equations are written for the two weighings and the density of the solid object is calculated from a solution of the two weighing equations.

In the simple ideal case, the following assumptions[2] are made:

1. A simple but perfectly linear spring scale (force-balance) is used as a detector.
2. The temperature of the air, liquid, and object are in mutual equilibrium.
3. The air density and the spring constant for the scale remain unchanged during the weighings.
4. The detector scale reads zero when the pan is empty both in air and in liquid.

WEIGHING EQUATIONS

The simple weighing equations are[2]

For weighing in air:

$$M_x g[1 - (\rho_a/\rho_x)] = KO_{AL} \qquad (21.1)$$

For weighing in liquid:

$$M_x g[1 - (\rho_L/\rho_x)] = KO_{LL} \qquad (21.2)$$

M_x is the mass of the object of density ρ_x; ρ_a is the density of the air; ρ_L is the density of the liquid; K is the spring constant; O_{AL} is the detector observation when the scale is loaded with the object in air; and O_{LL} is the detector observation when the scale is loaded with the object in the liquid.

When the two weighing equations are solved for ρ_x, the result is

$$\rho_x = (O_{LL}\rho_a - O_{AL}\rho_L)/(O_{LL} - O_{AL}) \qquad (21.3)$$

If an object S of known density, ρ_S, were weighed in air and in a liquid of unknown density, ρ_L, Equation 21.3 can be rearranged to solve for ρ_L. The resulting equation is

$$\rho_L = \rho_S[1 - (O_{LL}/O_{AL})] + \rho_a(O_{LL}/O_{AL}) \tag{21.4}$$

In the usual case, the simple weighing conditions listed above are replaced by those that permit variations in ambient conditions. Schoonover et al.[2] have shown how to determine the density of a liquid, ρ_L, by weighing a solid object density standard of density ρ_R in air and in the liquid.

If the weighings were done on an electronic balance calibrated using a built-in weight of density ρ_C at 20°C, the following more-complicated equation is used to infer the density of the liquid at the bath temperature:

$$\rho_{LBt} = [\rho_{Rn} \{[1 - (\rho_{aa}/\rho_C Z_1)]/[O_{ca}/(O_{2a} - O_{0a})]$$

$$- [1 - (\rho_{aL}/\rho_C Z_2)]/[O_{cL}/(O_{2L} - O_{0L})]\}$$

$$+ \rho_{aa} \, X \, [1 - (\rho_{aL}/\rho_C Z_2)]/[O_{cL}/(O_{2L} - O_{0L})]]/$$

$$Y \, [1 - (\rho_{aa}/\rho_C Z_1)]/[O_{ca}/(O_{2a} - O_{0a})] \tag{21.5}$$

where ρ_{LBt} = density of the test liquid at the bath temperature, Bt, ρ_{Rn} = density of the solid object density standard at 20°C, ρ_{aa} = air density during the weighing in air, ρ_C = density of the built-in weight at room temperature, $Z_1 = 1/[1 + 3\alpha(t_{aa} - 20)]$, ρ_{aL} = air density during the weighing in the liquid, $Z_2 = 1/[1 + 3\alpha(t_{aL} - 20)]$, $X = 1 + 3\beta(t_{aa} - 20)$, $Y = 1 + 3\beta(t_L - 20)$, O_{ca} = balance indication with calibration weight engaged for weighing in air, O_{2a} = balance indication when loaded with the solid object density standard in air, O_{0a} = balance "zero" reading (pan empty) for weighing in air, O_{cL} = balance indication with calibration weight engaged for weighing in the liquid, O_{2L} = balance indication when loaded with the solid object density standard in the liquid, O_{0L} = balance "zero" reading (pan empty) for weighing in the liquid, t_{aa} = air temperature for weighing in air, t_{aL} = air temperature for weighing in the liquid, t_L = temperature of the liquid, α = coefficient of linear thermal expansion of the built-in weight, and β = coefficient of linear thermal expansion of the solid object density standard.

DETERMINATION OF THE DENSITY OF LIQUIDS USING A MECHANICAL OSCILLATOR TECHNIQUE

The density of liquids can be determined very precisely and accurately using a device embodying a mechanical oscillator technique.

The sensing element of the device is a U-shaped, oscillating sample tube. The sample tube is completely filled with 0.7 milliliters of the liquid under test. The filling is done by injection using a hypodermic syringe or by other means. The sample tube is excited electromagnetically. The density of the liquid is determined from the measurement of the period of oscillation of the sample tube.[3]

The device is conventionally calibrated with air and a liquid of known density (e.g., doubly distilled water). The attainable precision of the device is strongly dependent on the temperature control of the device. For example, the claimed precision of the device is 1×10^{-4} g/cm^3 for thermostat precision of ±0.05°C, and 1×10^{-5} g/cm^3 or 1.5×10^{-6} g/cm^3 for thermostat precision of ±0.01°C.

REFERENCES

1. **Bowman, H. A. and Schoonover, R. M.,** Procedure for high precision density objects, *J. Res. Natl. Bur. Stand.,* 71C, 179, 1967.
2. **Schoonover, R. M., Hwang, M.-S., and Nater, R.,** The determination of density of mass standards; requirement and method, *NISTIR 5378,* 1994.
3. **Kratky, O., Leopold, H., and Stabinger, H.,** The determination of the partial specific volume of proteins by the mechanical oscillator technique, *Enzyme Structure,* Methods in Enzymology, Vol. 27D, Academic Press, New York, 1973, 48.

Chapter 22

TANK VOLUME

INTRODUCTION

Tanks are used for the collection of fluid that has passed through a flowmeter being calibrated and can be used to dispense liquid through a flowmeter for calibration.

As shown in Chapter 12, the differential pressure (DP) between two points or levels in a liquid depends on the density of the fluid (ρ), the acceleration due to gravity (g), and the vertical separation between the two points or levels. Therefore, in a simple system, if one could determine the DP between the bottom of a tank containing liquid and the surface of the liquid, the height of the liquid could be inferred from ρ and g. If the mean cross-sectional area of the tank were known or could be inferred, the volume of the liquid in the tank could be determined.

Changes in the height (and therefore volume) of the liquid could be determined from DP measurements before and after collection or discharge of liquid.

TANK CALIBRATION

The tank is calibrated by adding known increments of water to the tank and measuring the differential pressure. The increments of water can be introduced to the tank volumetrically from volumetric test measures (see Chapter 20) or increments could be weighed before introduction and the volume calculated using the density of water calculated from water temperature measurements (see Chapter 3).

The DP between the bottom of the tank and the air space above the liquid surface is determined by the use of two "bubbler tubes", one with an orifice near the bottom of the tank and the other with an orifice above the surface of the water. As in Chapter 12, air is forced through the bubbler tubes and bubbles form and break off the ends of the tubes. The pressure at the effective tip of each bubbler tube is transmitted to one side of a pressure-measuring device by the air. The DP across the device is indicated by the device.

CALIBRATION EQUATION

Calibration data, DP, and liquid volume (V), are fitted to a linear equation of the form:

$$DP = a + bV \tag{22.1}$$

where a and b are coefficients determined by fitting the data by linear least squares. In practice, Equation 22.1 is fitted to data in vertical segments in the tank.

157

Jones[1] has applied the tank volume calibration system to chemical process tanks and has developed a tank volume calibration algorithm[2] to permit inference of liquid volume in a tank from measurements of DP and temperature, and the values of other parameters. An American National Standards Institute (ANSI) standard, "American National Standard for Nuclear Materials Control—Volume Calibration Techniques", N15.19-1989,[3] is a detailed treatment that is generally applicable to any tank equipped with a system for measuring liquid content.

The method is applicable to water and to other liquids for which the density is known.

APPLICATION TO FLOW MEASUREMENT

Equation 22.1 can be inverted to enable calculation of liquid volume from determination of DP. The inverted equation can be applied to determine the quantity of liquid transferred into or out of the tank through a flowmeter.

REFERENCES

1. **Jones, F. E.,** Application of an improved calibration system to the calibration of accountability tanks, in *Nuclear Safeguards 1978*, Vol. 2, International Atomic Energy Agency, Vienna, 1979, 653.
2. **Jones, F. E.,** A tank volume calibration algorithm, *Nucl. Mater. Management*, Spring, 1984, 16.
3. American National Standard for Nuclear Material Control Volume Calibration Techniques, ANSI N15.19-1989, American National Standards Institute, New York, 1989.

INDEX

A

Air, 46
  density, 72
    calculation, 12, 17
    uncertainties in, 17
  -free water, 28, 30
    density of, 31–32, 35–36
    isothermal compressibility of, 27
  gamma, ratio of specific heats for, 57–63
    real-gas specific heats of dry air,
      57–59
    real-gas specific heats of water vapor,
      59–63
  nonideality of moist, 37
  precision estimates for, 49
  -saturated water, density of, 33–34
Analysis of variance (ANOVA), 104, 107
Anemometers,
  ascertaining linear range of, 135–138
    application of analysis, 137
    Jones' treatment of calibration data for
      turbine flowmeters, 135
    treatment of anemometer calibration
      data, 135–136
  linearizing output of, 137
ANOVA, see Analysis of variance
Apparent mass, 75, 79
Argon, 51
  abundance of, 13
  atomic weight of, 22
  calibration data for, 113
  in dry air, 11
  viscosity of, 41, 45, 54
Automatic pipets, 95–97

B

Background value, 11
Balance
  adjusting of, 76
  calibration of, 76
British Engineering, 3
Bubble
  tube, 101, 157
  radius of curvature of, 102
Buoyancy
  correction factor, 71, 73, 80
  force, 71

C

Calibration, 145
  data, 126
  factor, 130
  flow-related, 80
  tests, 134
Carbon dioxide, 49
  abundance of, 11, 17
  in dry air, 11
  mole fraction of, uncertainties in
    measurements of, 20
  viscosity of, 41, 44, 54
Coefficient of linear expansion, 93
Collection tank, 141
Combined standard uncertainty, 67
Compressibility factor, 10, 13, 15, 87
Constant pressure, 60
Containment volume, 149
Continuity, equation of, 85
Conversion equation, 28
Coverage factor, 67, 69
Critical pressure, 89
Critical-flow nozzles, 83

D

Delivery volume, 149
Detector scale, 153
Differential pressure (DP), 5, 100, 114, 157
Discharge coefficients, subsonic flow and,
  85–93
  derivation of discharge coefficient for
    orifice plates, 90–93
  nozzle discharge coefficient, 86–90
    derivation of expression for C_D, 87–88
    uncertainties in nozzle C_D, 88–90
  subsonic flow of gas through venturis,
    nozzles, and orifices, 85–86
Distilled water, 150
Diverter corrections, 139–144
Diverter valve correction, 141, 142, 143
DP, see Differential pressure
Dry air, 38
  apparent molecular weight of, 11, 12
  calibration data for, 113
  composition of, 11
  specific heats for, 57
  viscosity of, 41, 42, 54

159